Science Centres and Science Events

Anne-Marie Bruyas • Michaela Riccio
Editors

Science Centres and Science Events

A Science Communication Handbook

Springer

Editors
Anne-Marie Bruyas
International Relations
Fondazione IDIS-Città della Scienza
Naples
Italy

Michaela Riccio
International Relations
Fondazione IDIS-Città della Scienza
Naples
Italy

In cooperation with: FONDAZIONE IDIS CITTA'DELLASCIENZA

ISBN 978-88-470-2555-4 ISBN 978-88-470-2556-1 (eBook)
DOI 10.1007/978-88-470-2556-1
Springer Milan Heidelberg New York Dordrecht London

Library of Congress Control Number: 2012933797

Printed on acid-free paper

Springer is part of Springer Science+Business Media (www.springer.com)

Foreword

Science is all around us, it is part of our culture and has a great impact as such on our lives. New scientific advances – and indeed new technologies based on existing knowledge – are allowing our lifestyles to change at a faster rate than ever before. Today, more than ever, we are relying on science and scientists to help us answer questions and address problems that were unthinkable even just few years ago.

What should be communicated and how? What does actually involve the concept of scientific citizenship, which gives a democratic form to knowledge society? And how is it possible to construct it? How to build science events? These are some of the questions addressed by this Handbook. The publication includes chapters written by leading science communicators and experts in science centres, each providing a comprehensive review of important current issues in science centres, educational activities and science events. As the editors explain, the book places the Handbook within the context of the national and international activities of Fondazione Idis – Città della Scienza, a Science Centre which has been associated with some challenging UNESCO's events.

The Science Communication Handbook is addressed to future scientific communicators, or, in other words, to those who have chosen (or accepted) the extraordinarily difficult task of disseminating scientific information. It is a difficult task by its very nature, since it means making the fundamentals of science comprehensible and accessible to non-specialists, without undue reductionism or over-simplification. In the last few decades, what was previously a difficult task has in fact become a virtually impossible one, because the increasingly rapid pace of scientific and technological discoveries and inventions, with the broadening of the scope of knowledge and the narrowing down of specialist fields, even the fundamentals now seem to be beyond the grasp of the layperson.

As a Director of the Division of Science Policy and Sustainable Development at UNESCO, I am very happy to have been given this opportunity to write the foreword for this Handbook, since I consider science communication as an integral part of inclusive, people-centred science policy design. In this regard Science Communication represents one of the major tools to mobilize and to empower society, a society where people are intrigued, inspired and involved with science.

Science Communication becomes also an instrument for social transformation. In any democratic society, citizens have the right to influence decisions that are taken on a wide variety of matters on which they have an interest. Science is a matter on which the citizens have an interest. If the general public wants to have a voice in the scientific decision-making, helping to decide about these issues, then it must know better science. In other words, a science literate public is an empowered public. Today, more than ever before, access and participation to decision-making is not only a possibility to be explored but a moral necessity.

In this perspective science centres, by reflecting communication as an interactive process, and looking at science and technology in a larger context, are becoming more and more an "agora" for dialogue and interaction. Therefore, they become a public forum for debate, ongoing conversation and learning. In other words, a place where you can find the news and where transformations and changes might happen.

At the same time, science centres, abandoning a purely didactic and intellectual approach, become places of interactive culture, restoring the links between information and sensation, 'reverie', excitement, pleasure, intellectual provocation, curiosity and imagination. In the world of Albert Einstein, "Logic will get you from A to B. Imagination will take you everywhere".

Last but not least I wish to highlight and commend the practical experiences presented in the Handbook on how science represents a fundamental mean for intercultural dialogue, reconciliation and peace.

I would like to wish those using this Handbook an enjoyable experience in inspiring young people with science and science careers, in supporting teachers to encourage their students with STEM, in engaging the public with science and in widening participation in science though Science Communication.

Lidia Brito
Science Policy and Sustainable Development,
UNESCO

Acknowledgments

We would like to thank all who have contributed to the development and success of the training course of nine Nigerian university students, held in Naples from the 3rd to the 18th of September 2008, from which we got a great human experience of knowledge and the inspiration to turn this experience in a practical handbook.

First of all, we would like to thank Card. Gianfranco Ravasi President of the Pontifical Council for Culture and Mgr. Melchor Sanchez de Toca y Alameda, Under-Secretary for Intellectual Charity, for the institutional patronage to the initiative. Our warm acknowledgments go also to Rev. Fr. Tobechi Anyadike – father of the project for the establishment of a science centre in Owerri (Nigeria) and President of the ASCO Association (Assumpta Science Center Owerri) – who took care of the Nigerian students during their stay in Italy, together with Laura, Alessandro, Sabino and all the Italian students from the University of Rome "La Sapienza". A sincere thank to the supporting organisations who make this initiative possible: Istituto Banco Napoli-Fondazione and ICS (Istituto per il Credito Sportivo) for the training course in Naples, Finmeccanica and Fondazione IBM Italia for the Owerri Science Festival.

We would like also to thank all our colleagues of Fondazione Idis – Città della Scienza, who were directly and indirectly involved in the organization of the training course and of the Science Festival in Owerri, for their professionalism in delivering expertise but also for their warmth and friendship that made young Nigerian felt at home. A special thank goes to Arduino Mangoni who worked with great enthusiasm with us to the coordination of the project.

Nobody among the Nigerian students will forget the few Italian words learnt with Franca and Rosa during the evening Italian language lessons, often burst out in a collective laughing. We warmly thank Marilù Vacca, Franca Sibillo, Rosa Giordano from CEICC, the Naples Centre for European Citizenship, for the time and the creativity they deployed in teaching Italian to them.

Isa Rampone, Giulia Gaeta Marmo and Federico Brigida, by teaching chemical demonstrations to Nigerian students during the training, made possible the establishment of the Chemistry Bar as an officially part of the Science Festival in Owerri.

Nigerian students spent one day in the amazing hands-on exhibition 'The Square Wheels' in Calitri, where Pietro Cerreta and his team showed them fascinating science phenomena and the secrets of handicrafts in mounting hands-on exhibits. We thank all the team of Scienza Viva Association for the wonderful hospitality they reserved to Nigerian students in the green Irpinia landscapes.

Finally our deepest thanks go to Placid, Declan, Boniface, Sylvia, Vivian, Charity, Genevive and Ngozi who spent two weeks in Italy learning expertise and techniques in science communication, and who went back to Nigeria with a great enthusiasm to organize the first Owerri Science Festival. They pursue today their efforts to organize activities in the temporary exhibition centre and in rural areas. We hope faithfully that they will be the future makers of a permanent Science Centre in Nigeria.

Anne-Marie Bruyas
Michaela Riccio

Contents

Introduction

Anne-Marie Bruyas and Michaela Riccio

A Handbook: Who Is Being Addressed

Between July 2008 and May 2009, we carried out a project for the development of a science centre in Owerri, Imo State, Nigeria. After this extraordinary experience and other international cooperation programs such as the establishment of a new science centre at the Palestinian Al Quds University in East Jerusalem, what we have obtained in the field has led to International recognition and has prompted us to prepare this text in order to provide a theoretical and practical tool for future scientific communicators who wish to become involved in setting up new projects.

The publication is designed to be a practical manual for those intending to use scientific communication as a tool for promoting sustainable development. In particular, we hope that it can be used as a training aid for developing countries where actions of this type are perceived to be lacking in priority or urgency.

One of the aims is therefore to raise awareness among associations and local cultural actors regarding both the significance of scientific and technological culture for the development of underprivileged areas as well as the strategic role of scientific communication for achieving such development.

In these countries, the role of scientific education and training in schools is not sufficient to arouse scientific curiosity among young people and make the population aware of the importance of scientific knowledge in everyday life. Science and technology are indispensable tools for people's empowerment and should be supported with actions that encourage curiosity about science and the intelligent use of technology to bridge the divide with developed countries. It is therefore necessary to set up activities that are carefully targeted to promote and communicate science (Fig. 1).

A.-M. Bruyas (✉) • M. Riccio
International Projects and Relations, Fondazione Idis-Città della Scienza, Naples, Italy
e-mail: bruyas@cittadellascienza.it; ricciom@cittadellascienza.it

A.-M. Bruyas and M. Riccio (eds.), *Science Centres and Science Events*,
DOI 10.1007/978-88-470-2556-1_1, © Springer-Verlag Italia 2013

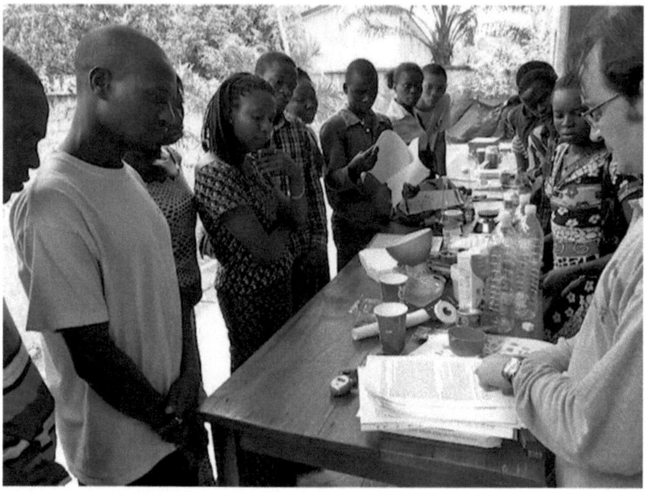

Fig. 1 Training of local staff at the Owerri Science Festival, Nigeria

For these reasons the publication has been designed as a practical book to be used in different types of events: scientific events or more structured science festivals, training contexts, the creation of scientific cultural associations, or the development of new science centres.

Besides being an excellent tool for training and supporting the design and planning phases, this manual can also be used as a reference work for institutions and local cultural services which have to choose and select projects of this type.

What Is Città Della Scienza?

Fondazione Idis-Città della Scienza is a community that lives and works in an old chemical factory, reconstructed by a wonderful project of industrial archaeology, facing the bay of Naples, under the hill of Posillipo, between Nisida and Bagnoli (Fig. 2).

Lying over 65,000 m^2, there is a modern interpretation of a concept – rather a value – which is one of the never written but in function, basic foundations of the "Science Republic": communicate everything to everyone, to make scientific knowledge a good available not only to a privileged few, but to everyone. The Città della Scienza – created by Vittorio Silvestrini – doesn't only host the first hands-on museum in Italy; but also a place where public understanding of science has become the lever for new development. The story of Città della Scienza isn't only the success story of the diffusion of scientific knowledge, but also the story of the construction of a new development model: a cultural and economic model that come from the decline and enter into the society of knowledge (Fig. 3).

Fig. 2 Città della Scienza and Posillipo hill

Fig. 3 General view of exhibitions, Città della Scienza

Claiming a new scientific citizenship, bridging the gap between scientists and society at a whole, for a shared control of science and technology, in order to create the conditions for a dialogue with ordinary citizens, is one of the primary goals of Città della Scienza. Furthermore, the training of new generations and the transmission of a scientific method is essential to the development of a dynamic society open to the challenges of the future. Among its several public programs, Città della Scienza engages and stimulates the interest of young people and their creativity, making them aware of the issues of current scientific works by bringing them closer to scientists and researchers, inspiring in such way new science careers (Fig. 4).

In a time of globalization, scientific disciplines own also a key role in the achievement of a better quality of life and in the affirmation of a knowledge-based

Fig. 4 Hands-on exhibit on fluid dynamics

economy. The access to resources of knowledge, thanks largely to the spread of new communication technologies, is an essential factor for economic growth. Innovation is considered as a search for better and more efficient solutions, but moreover it should be considered as a "new area of potential forces and resources." It means the right – to be protected and guarantee for the whole community – to share the benefits of a more open, integrated and cohesive society. Through its activities, Città della Scienza contributes to design a restructuring of the economy for compatible sustainability and competitiveness, social as well as ecological. Città della Scienza won in 2005, under the auspices of the Council of Europe, the *Micheletti Award* for the best science museum in Europe. Its founder Vittorio Silvestrini was awarded in 2007 by the European Commission with the *"Descartes Prize"* which recompenses excellence in scientific communication. While in 2007 and 2008 Città della Scienza won the *Best Science Based Incubator Award.*

Fondazione Idis-Città della Scienza has always focused its activities on the European and Mediterranean context, realizing exchanges as well as scientific and cultural cooperation programs between European regions and the Southern Mediterranean Countries, promoting transnational programs and projects, and realizing activities that state science as a universal language of peace.

Città della Scienza is promoter of the Euro-Mediterranean and Black Sea Observatory, together with the Municipality of Naples and the University of Naples, a partnership that promotes projects and activities of decentralized cooperation based on the values of peace, development and solidarity.

Città della Scienza is also a project leader for the creation of innovative science programs in crisis areas and developing countries, such as Palestine, Iraq and Nigeria, believing that the spreading of scientific culture, especially to youngest generations, represents the *conditio sine qua non* to the progress of such regions.

Città della Scienza looks also to emerging countries (China and East Asia) where it operates through developing programs of scientific and technological exchanges with the areas of Beijing and Zhejiang (China), supporting courses and activities for the internationalization of the innovative regional system (companies, research centres, universities) of those areas.

Finally, Città della Scienza is strongly involved in International professional networks: ECSITE; the European network of science centres and museums, with Vincenzo Lipardi being President from 2007 to 2009; the International programme Committee of the Science Centre World Congress; and NAMES the North African and Middle East Science Centre, launching in 2012 a professional training Summer School for science communicators in the Mediterranean.

Part I
Science in Society

Toward Knowledge Societies

Giuseppe Vittorio Silvestrini

Sixty years have gone by since the father of cybernetic science, Norbert Wiener, heralded the advent of a new era of information and knowledge which he predicted would bring about enormous opportunities and risks. Nowadays nearly everyone recognizes that the world has entered a new season of cultural and economic evolution, based on science and on a type of technology which, to use the words of the Italian sociologist Luciano Gallino, "incorporates endlessly growing volumes of scientific knowledge".

We have entered an era in which the engine of society and the economy of mankind no longer consist of physical and thus limited resources but of an "infinite resource": science and, more generally, knowledge. Knowledge society represents both the development and the downfall of industrial society, based on machinery and labour, which in turn represented the development and the downfall of rural society, based on the domestication of plants and animals and on manual labour in the fields.

The radically new aspect of our own era is clearly not knowledge in itself. The cultural evolution of mankind began with Man himself. The production of new knowledge, information and innovative technologies has always accompanied mankind. Indeed, anthropologists have coined the term *habilis* to refer to the first representative of the genus *Homo* who appeared on the Earth over 2.5 million years ago. And what does the term *habilis* refer to but the capacity to acquire new information, transmit it rapidly and transform it into knowledge and use it to manipulate the surrounding environment through labour and technology? The novelty of our era lies in the fact that the production of new information, new knowledge and the type of technologies that "incorporate endlessly growing volumes of scientific knowledge" have become the primary factors, to quote Luciano Gallino once more, "of innovation, economic growth, and international

G.V. Silvestrini (✉)
President of Fondazione IDIS-Città della Scienza, Naples, Italy
e-mail: silves@cittadellascienza.it

A.-M. Bruyas and M. Riccio (eds.), *Science Centres and Science Events*,
DOI 10.1007/978-88-470-2556-1_2, © Springer-Verlag Italia 2013

competitiveness of firms and the national economy". In other words, the production of goods with the greatest market success in information and knowledge society increasingly requires a lower intensity of labour and a higher intensity of knowledge.

Actually, the construction of the knowledge society is a long process that has involved several stages. Even in the late nineteenth century, for example, and to an increasing extent during the twentieth century, the evolutionary process of human civilization has undergone a brusque and irresistible acceleration. The main cause has been the systematic adoption – especially by western countries – of the scientific method as a tool for increasing knowledge. Full-scale cultural revolutions followed and overlapped one another; each generation has somehow managed to achieve a far-reaching transformation of knowledge, processes, values and goods, bequeathing its children and grandchildren a civilization and world that was profoundly altered compared to the ones they had inherited. The huge conquests in knowledge (thermodynamics, electromagnetism, material sciences, nanosciences) are full of revolutionary applications and consequences (heat engines and electric engines, electronics, new communication and information technologies and nanotechnologies). They have provided human civilization with an enormous positive potential in terms of the productive capacity of goods, tools and services which is capable of bringing about great improvements in the quality of life but is also full of dangers if used without necessary discernment.

It is therefore worth recalling some of the key moments in this rapid and turbulent phase:

- The so-called *industrial revolution*, based on the spread of the productive system of machinery capable of assisting man and increasing operational capacities, especially in the manufacturing sectors. As a result of this revolution (accompanied and reinforced by the *electronic revolution*, which has gradually improved the productive capacity of machinery, equipping it with an increasingly sophisticated and powerful brain), the capacity of industrialised countries to produce wealth and improve the quality of life of humanity has dramatically increased. During the first half of the twentieth century, all families acquired the real possibility of having their own car, fridge, radio and later a television etc. Meanwhile, the levers of decision-making power and the flows of increasing added value passed into other hands, leaving the working classes and becoming concentrated in the hands of those who owned the capital required to provide and control the means of production.
- The *ICT revolution* (ICT, Information and Communication Technology), as a result of which there has been an increasing capacity to produce and spread a steadily larger amount of information, offering each citizen the real possibility of participating in this immaterial banquet, gaining nourishment and enjoying the opportunity to contribute and enrich supply. At the same time, it provided economic players with more powerful tools for influencing the tastes and convictions of collective society in order to make them tools of their own economic, political and cultural interests etc.

- The *globalisation of markets*, created by the explosion of ICT. The opportunity to transfer information, instructions and decisions to every corner of the world in real time makes it possible to organise the productive system by distributing moments and structures throughout the world according to expediency and cost. The uniformity of tastes and needs, accentuated by the scanning capacity transmitted by the mass media, transforms the productive system into a formidable machine for producing wealth: this wealth derives from thousands of trickles which move inwards from the periphery to the centre, leaving those who are excluded and marginal with just the bare crumbs to survive, and quite often not even those.

The knowledge society therefore offers enormous opportunities as well as huge risks. This is proven by the fact that never before, despite the recent crisis, has mankind produced such wealth. But never before has there been such inequality. Knowledge does not benefit everyone but only a small minority. The knowledge society, as we have known it until now, is socially unsustainable.

In the knowledge economy, the burden of processes, goods and immaterial services continue to grow. And yet the human imprint on the local and global environment has never been more profound and invasive. The knowledge society as we have known it until now is ecologically unsustainable.

However, a different world is still possible. And this is possible not by giving up knowledge but by democratising knowledge, by making it available to the many rather than just the few. It is neither possible nor opportune to prevent innovation of the system of supplying goods. On the other hand, it is true that it is opportune, and indeed necessary and urgent, to intervene in the system of demand – in the direction of the demand for public goods – in order to reinforce it so that it can express the needs linked to 'being' (the needs of the quality of life) rather than appearance (the accumulation of private goods). It is also important that the system of demand can find an answer to these needs, even when the answer is not provided by the canonical system of supply. In other words, it is necessary to rediscover and reinforce the economy of use values (values that can take account of the utility of a good and the human work required to produce it) by reinventing new mechanisms in symbiosis with the system of supply. Meanwhile, however, it is necessary to find new "parallel spaces" for the supply system with respect to existing spaces; these spaces are currently occupied in an invasive manner by an economy based on exchange values (where goods are not valued on the basis of their utility but on the basis of conventional parameters that take account of the credibility of those that possess them), together with an economy based on the production of goods of symbolic value (goods that are designed for appearance rather than 'being' and which promote consumption and waste).

We should therefore avoid rejecting scientific knowledge and innovation: on the contrary, we should take full possession of this resource and put it to the use both of supply and demand in order to sustain the development of an economy of use values that fills the "vacuums" created by the dynamics of globalisation.

For example, we could launch the production of local and global "public goods". In other words, we can set up activities designed to produce and promote not just consumer goods but more importantly what can be termed "the details of the quality of life". By way of example (although this is more than just a simple example), we could mention activities aimed at preserving the city and, more generally, at making it liveable, or activities aimed at guaranteeing citizens' health (with particular attention focused on prevention prior to actual treatment of illnesses), activities for protecting and preserving the natural environment and the local area, and educational activities, with special attention focused on the constructive use of free time. These are all socially valuable activities. They don't just provide answers to the basic needs of individuals and collective society but they are also marked by a high content of real work, as well as being strongly rooted in the local area and its resources and are thus difficult to expropriate. Generally they tend to produce widely distributed sources of income and more widespread employment, thus providing an answer to a social need that is particularly strongly felt – employment – to which the economy of global competition tends to provide an increasingly weak response. Nevertheless, the development of these activities will be far from automatic since the process is hindered by several obstacles that must be removed.

Firstly, there are obstacles of a cultural nature, both in terms of supply and demand. We live in a historical moment in which thousands of stimuli tend to lead to the fragmentation of interests, competition, defence of privileges, cunning of individuals to the detriment of the collective, consumption and waste rather than conservation; it is therefore becoming increasingly important to promote the culture of collaborative commitment; this is necessary to create the environmental conditions in which, even at the cost of sacrifices on the part of individuals, the life of the collective becomes more satisfying. In order to provide a response to this demand, once it has emerged, it is necessary to organise a form of supply that involves the rescuing of traditional knowledge and the grafting of the "know-how" offered by modern science onto such knowledge. It is therefore particularly important (though not sufficient!) that scientific knowledge and the potential of technology, which nowadays permeate the economy and civilisation especially in terms of supply (in particular due to the system of "strong" industry), manage to penetrate society at all levels beginning from demand. This requires a systematic process of mass scientific literacy, the organic metabolisation of new forms of knowledge within "classic" culture, and social participation in decisions concerning scientific policy in order to promote local systems of innovation.

There are obstacles of an economic nature and, as a consequence, of an organisational and political nature. It should be observed that even when the need for environmental conditions that can ensure a reasonable quality of life is widely felt, and there is a potential demand for activities linked to use values, this demand is unlikely to take on an economically viable form. Indeed there is no realistic prospect that each individual can aspire to buying his/her own private niche of environmental quality. It is therefore necessary to plan moments of social and political organisation that make it possible to assemble some sort of collective patron and to identify mechanisms designed to ensure that these bodies can procure sufficient resources to act as solvent patrons.

Given that the main incentive for these sectors is neither accumulation nor profit, but a demand for a quality of life that stems from the bottom-up, an important role in their organisation and running can be played by associations and "non-profit" organisations which belong to the "third sector". A fundamental role needs to be played by the State which should, at the very least, have the task of providing the legislative framework and a series of vital economic, fiscal and organisational incentives. However, important spaces are being created through local firms whose entrepreneurial mission is represented by the enhancement of environmental and collective resources (heritage, artistic and cultural resources, cultural traditions, craft goods and local food and produce etc.) in order to meet local demand, but also to meet remote demand and attract international tourism.

Only if we can organise a widespread demand for "use value" will be finally be able to grasp all the opportunities provided by the development of scientific research and ensuing technological innovation. In this way, we shall finally be able to build a democratic knowledge society.

nature and to identify mechanisms designed to ensure that these bodies can promote [illegible] responses to [illegible] problems.

[illegible]

Science and Society

Pietro Greco

The World Has Changed

The world is not changing. The world has already changed. The political geography of the world has changed. Only 20 years ago, the Soviet Union still existed and the world was divided into two blocks which were opposed each other ideologically as well as militarily and economically. Many non-aligned and largely uninfluential countries were merely spectators of this intense cold war between two blocks that were armed to the teeth. Nowadays there is a single superpower, the United States, which is involved in a series of conflicts of a more minor nature and/or lesser intensity, while a series of countries with developing economies are acquiring new importance and new roles.

The economic scenario has also changed, and just as radically, in three important ways:

- The world has become wealthier: in the last 20 years the wealth of the planet has doubled. Before the recent crisis, economic growth had grown constantly for 40 years; in the 1970s the GDP of the planet accelerated at an average rate of just over 3% a year, in the 1990s the increase in wealth produced by the planet exceeded on average 4% a year and in 2006 the growth in world GDP stabilised at about 5%. The increase in GDP per capita, the average wealth (only average, alas) that each citizen of the planet possesses rose sharply: it grew at an annual rate of 2% in around 1970 and increased at a rate of 3% in the first few years of the new century.
- The places of wealth production have changed: in 1985 Europe was still the continent with the highest production level (25% of global GDP); today Asia has the highest levels of production (36% of global GDP). Besides the so-called advanced free market countries (USA, Japan and European countries), the

P. Greco (✉)
Science writer and founding member of Fondazione Idis-Città della Scienza, Naples, Italy
e-mail: pietrogreco011@gmail.com

A.-M. Bruyas and M. Riccio (eds.), *Science Centres and Science Events*,
DOI 10.1007/978-88-470-2556-1_3, © Springer-Verlag Italia 2013

developing economies of China, India, Brazil, Russia and a whole series of countries situated mainly in south-east Asia have become the protagonists of an increasingly globalised market. Indeed, these are the countries with the highest rates of growth. For example, the Chinese economy has been growing at an incredibly rapid rate of over 10% a year.
- However, what has really changed are the means of wealth production: we have now become a knowledge society. The most dynamic firms, both in the manufacturing and service sectors, are those that produce and apply new forms of knowledge. They create more wealth than other firms, provide better remuneration for work and encourage the general growth of the economy to a greater extent.

All this is also linked – arguably chiefly linked – to the role of science and technological innovation, based on science, in economy. For at least two decades, the new dimension assigned to scientific research and technological development (R&D) in a large part of the planet has followed the change in the global economy. In order to boost the growth and competitiveness of their countries, many governments have encouraged the development of a knowledge-intensive economy, thanks to generous incentives to scientific and technological research, the development of infrastructure, the extension of access to higher education, as well as the introduction of new regulations. These policies have led to the development of a market economy that is increasingly based on knowledge.

There are at least three crucial factors for change that are currently taking place in the *science-intensive economy*:

- Growth in investment in R&D. In the last 20 years global investment in R&D has more than tripled, taking into account inflation, rising from just over 350 billion dollars in 1990 to over 1,100 billion dollars in 2008; this is a figure that has never been spent before and now equates to 2.1% of global wealth.
- Growth in private investment. The second structural statistic is that investment in R&D by private firms has increased at a much higher rate than public investment. Worldwide, for every public dollar spent on R&D, two private dollars are spent.
- The geography of research has changed. We are rapidly moving from a bipolar world of science with research carried out only in Europe and North America (with the exception of Japan) towards a multi-polar world. New countries are deploying imposing economic and human resources: China, India, a whole set of countries in south east Asia, but also Brazil, Mexico, and South Africa, are becoming the protagonists of the knowledge society.

Science and State

In July 1945, while the war had ended in Europe and the war in the Pacific was still continuing, Dr. Vannevar Bush, the director of the US *Office of Scientific Research and Development*, wrote a report for the new US President Harry S. Truman, entitled *Science: The Endless Frontier*.

Vannevar Bush, who had been the influential scientific advisor to Franklin D. Roosevelt, was coordinating American researchers who had been mobilised to win the war; these physicists were bringing the Manhattan Project to a conclusion. The report he compiled for the newly elected President Truman was destined to mark a major turning point in the history of the "Republic of Science". This was because Dr Bush identified basic science carried out in universities as the cornerstone for building the national security of a power that was about to win the Second World War and plan a New World Order for the period of peace that was supposed to follow.

As soon as the war was over and the time of peace had begun, the author of the report argued, *the federal government should immediately dismantle the Office of Scientific Research and Development*, directed by Bush himself; it had been set up for wartime and had mobilised thousands of academic scientists with extraordinary results. Nevertheless, the report continued, it was necessary for the federal administration to preserve the spirit of the Office and to continue to mobilise academic scientists to obtain equally extraordinary results which will enable the American nation to meet the challenges of the post-war period and consolidate the economic, social and health security – as well as military security – of the United Sates. Mobilising academic science means bestowing on it great generosity, financial and human resources in order that science can conduct its research in complete liberty, even when it involves studies – like pure mathematics or astrophysics – that have no immediate practical spin-offs.

After the report – and after the explosions of Hiroshima and Nagasaki offered the whole world a tragic but powerful demonstration of the results to which the mobilisation of academic scientists can lead – the personal fortunes of Dr Bush began to decline. However, many of his suggestions would be followed by Truman and his successors in the White House. Science took on a strategic role in the United States both in a military context but also in a civilian context. New organisations were set up such as the *National Science Foundation*, which was given the task of supporting and coordinating research activities in American universities by the federal government and the *National Institutes of Health*, entrusted with supporting and coordinating biomedical research activities. Subsequently NASA would be set up and would bring the American dream into space. Due to the precise political decision based on the advice of Vannevar Bush, science became the main lever of American development and the scientists that worked in universities found themselves managing unprecedented financial and human resources in all sectors of research, both pure and applied. It was a strategic and purely political decision that has enabled the United States over the last 60 years to become the most powerful military power and the leading country in the planet for technological and economic development.

The success of the strategy outlined by Bush – academic science as a lever for development – caused such a radical evolution in the relationship between science and society that his report *Science: The Endless Frontier* can rightly be considered to be the start of what John Ziman called the transition from academic science to post-academic science; this transition radically redrew the physiognomy of the scientific world through new relationships between research and politics.

In its earliest stages in the seventeenth century, the scientific world had claimed autonomy from both political and religious power. The relationships between scientists and politicians (or, more generally, between scientists and power) were neither rare nor detached – the seminal figure being Galileo Galilei who in 1609 offered his perfected telescope to the Senate of the Republic of Venice and in the spring of 1610 dedicated his *Sidereus Nuncius* to Cosimo II de' Medici, Grand Duke of Tuscany. However, despite this and despite the fact that patronage, the generous benevolence of the powerful, was one of the main sources of support for mathematicians, astronomers and natural philosophers throughout Europe, the practice of autonomy by the new fledgling transnational community was not encroached on. This was because ultimately science appeared to be more of a cultural endeavour which had a value in its own right rather than the lever of a nation's economic and military development. For over three and a half centuries this situation remained unchanged despite the evolution of political regimes, the development of scientific knowledge – including immediately applicable knowledge – and the structural changes in scientists' method of working.

Only in the nineteenth century the profession of scientist became fully recognised and the careers of people engaged in research began to take place in universities. It was only in the nineteenth century that "academic science" was established, formalising the relationships between the community of scientists and the political institutions of single states.

Despite this transition, the community of scientists – which had become an academic community – continued to regulate itself, according to the sociologist Robert K. Merton, on the basis of five fundamental original and shared values. The values are known with the acronym CUDOS.

C	Communalism	Public knowledge of the results of scientific discoveries and public access to the results of research
U	Universalism	Anyone can provide a contribution to the progress of knowledge independently of their race, nationality, gender, religious beliefs, or political persuasion
D	Disinterestedness	By providing their contribution to the development of knowledge, scientists are not pursuing their own personal interests
O	Originality	The production of new and original knowledge
S	Organised Scepticism	In the scientific community, the notion *Ipse dixit* has no validity since everything can and must be open to discussion and verification

Naturally, research was not just carried out in universities. As early as the nineteenth century and to an even greater extent at the beginning of the twentieth century, groups of scientists worked in and for industries or for specific state objectives. Moreover, scientific knowledge began to be used systematically for producing technical innovation with a consequently high social impact. Nevertheless, during this period, the relationships between science and politics and between science and society were not organic and did not involve the entire scientific community: academic scientists did not receive a salary for doing research. They were employed in universities to teach and train technicians and were entitled to carry out research when they were not doing any teaching work. In this (formally) secondary activity, political institutions did not intervene in an organic or systematic way either to monitor or to give guidelines for research to the entire scientific community. Given this type of state patronage, most of the crucial decisions for the development of knowledge – especially basic knowledge far removed from immediate applications – were taken independently within the scientific community itself.

After the turning point in the relationship between science and politics following the Second World War, four significant steps took place in America, defining together a new transition.

The first was of a quantitative nature: the federal government began to fund research with large sums of money, equal to several percentage points of national wealth. In 1930 the United States invested the modern-day equivalent of 1.5 billion dollars in research and development. Twenty years later, the USA invested the equivalent of 30 billion dollars, 20 times the previous sum. Today they invest roughly 360 billion dollars, 11 times the amount invested in the 1950s and 240 times the amount invested in 1930. This led both to a rapid increase in the number of people involved in scientific research (there are more scientists working in laboratories today than the entire number of those that lived in previous eras), and in the planning and implementation of studies and projects – for example, the landing of man on the moon – which had previously been inconceivable.

The second was of a qualitative nature: the US federal government not only financed applied, civic and military research but also provided considerable funding for basic research or, to use the current term, curiosity-driven research. A significant share of public investment in research and development (about 20%) was destined to purely academic science which did not have immediate practical applications (or at least this was not the aim).

The third step regards scientists' method of working. In some sectors, such as that of experimental physics, research began to be carried out by increasingly larger groups and in ways that depended increasingly on large-scale equipment and substantial funding, which could not be decided in a university context. This led to the establishment of *big science* which had had its secret prototype in the wartime Manhattan Project. The large-scale projects – from the moon landing to research into new particles with large accelerators to the Genome Project – involved thousands of scientists and billions of dollars.

The fourth new aspect regarded the decisive fact that the generous funding granted to the scientific community had political motives, even taking into account the need for methodological autonomy in research. The funding was debated and decided by politicians in government and Congress in Washington.

A largely similar process, although different in form, spread rapidly throughout the rest of the industrialised world. In the West (Europe, Japan) and in Communist block countries (the Soviet Union in particular) science took on a strategic role, both in the military and the socio-economic sphere. States made it their task to define a coherent "research policy".

During the post-war period there was a new major development. The general separation between (academic) science and politics was replaced by a dense and growing interpenetration between the two, since those states with fully developed economies started to look at research as a strategic resource, in accordance with the predictions already made by the great economist Joseph Schumpeter in the 1930s.

The old ivory tower opened up its gates and the relationship between science and society started to become much closer. In exchange for unprecedented resources, the scientific community was therefore forced to take important decisions for the development of research in participation with a series of non-experts: politicians, bureaucrats, managers and the general public. It was at this point that researchers felt a pressing need to obtain social consensus for their work: the scientific community had to tackle the issue of "social responsibility of science" and to communicate science to the public (to the various types of public) of non-experts, while striving to maintain the independence from politics that had been one of the cornerstones of the "Republic of Science" in the seventeenth century.

On the other hand, society as a whole had become increasingly permeated by scientific culture and its technological spin-offs and therefore felt it necessary to control science by guiding its development.

As from the mid-twentieth century, the dialectic between science and society therefore became one of the founding elements of democracy in free societies (the situation being very different in Communist countries). In other words, it became a political factor. This led to some unprecedented consequences: both politicians and citizens asked to participate with the "government of techno-science" and to take significant decisions regarding their development.

In this novel relationship three key aspects emerged:

- In order to continue their research, scientists "must" seek a difficult social consensus both in political circles and public opinion.
- Politicians are required to make important decisions – like allocating resources – for the development of science but according to time-scales, forms and even values that differ from those in use among the scientific community.
- Lastly, all citizens experience the invasion of science (and techno-science) in their everyday lives and this forces them to make continuous choices, which ideally are well-informed and carefully pondered. Public opinion feels a new need – albeit in a confused but impelling way – to "know about science", both in order to carry out their political function and to be able to be masters of their own future.

At this point we could venture a definition of the ideal triangle for the government of society in the post-academic era of science: researchers who are willing to open the gates and windows of their ivory towers, politicians who recognise the importance of science and the complexity of regulating a techno-scientific society, and citizens who recognise the primary value of scientific culture for the civic and economic development of society. Except that...

Science and Markets

Except that in the relationship between science and politics, in the last quarter of the twentieth century, a third element has forced its way into the equation: private research. Private research further modified the reference framework, making it necessary to reformulate it once more.

This latest turning point also took place in the United States when, between the 1970s and the 1980s, research funding from private industry overtook federal funding for the first time in the post-war period. The main source of research funding ceased to be the State and became industry with its highly specific interests.

The structural transformation of the economy of American research was accompanied and indeed encouraged by precise political and cultural decisions, like the three which, by sheer coincidence, were taken in 1980.

- In 1980 the Patent and Trademark Office (PTO), after 9 years of reflection, granted Ananda Mohan Chakrabarty, biologist of the company General Electric, the patent to protect the intellectual property of a genetically modified bacteria species (called "multi-plasmid hydrocarbon-degrading *Pseudomonas*") that could digest hydrocarbons found in oil spills and industrial waste.
- In 1980, at the request of Stanford University, the Supreme Court recognised the right of intellectual protection for the technique of cloning recombinant DNA pioneered by Stanley Cohen and Herbert Boyer in 1973. The technique was of a general nature because it made it possible to carry out molecular analysis of the DNA not just of viruses and bacteria but also of plants and animals and was to become a key tool in biotechnology laboratories.
- Lastly, again in the year 1980, the United States Congress passed the Bayh-Dole Act, a law that encouraged public research centres to patent their inventions to obtain royalties with which to fund further research. Universities also started to patent their research and sell the rights to industries.

This series of laws encouraged and speeded up the invasion of firms in the spheres of science and technological innovation. In the 1960s two public dollars were spent on research and development in the United States for every private dollar. The situation has now been turned on its head: for every public dollar, two private dollars are spent. Throughout the world (with the sole but important exception of Italy) two thirds of research funding comes from private firms and companies.

The world economy was transformed because it increasingly began to be marked by the production of high technology goods created in the industries with the highest intensity of research and development. According to the Global Insight World Industry Service, the sales of hi-tech products in the world increased (taking into account inflation) between 1980 and 2003 at an annual rate of 6.4%, almost three times that of other products (2.4%). In 1980 the five industrial sectors represented 8.1% of the total of goods produced while in 2003 their share in world markets had more than doubled and amounted to 17.7%.

However, the series of laws passed in 1980 not only encouraged and accelerated the growth of the global economy but also the market-oriented culture in the world of research which, until then, had been based on a public-spirited culture as well as on public funding. New types of scientists emerged, who were both researchers and entrepreneurs. A new type of "entrepreneurial science" emerged; it was particularly active in the computing and biotechnology fields and once again remodelled– and this time on a global scale – the equilibrium within the science world.

University science and industrial science, which were now even further intertwined with each other, modified the old values underlying scientific research. The old Mertonian values of CUDOS became intertwined, argues John Ziman, with the new industrial values of PLACE.

	Mertonian values		*The new values*	
C	Communalism	P	Proprietary	States property rights
U	Universalism	L	Local	Attention to specific problems
D	Disinterestedness	A	Authority	Research conducted under managerial authority
O	Originality	C	Commissioned	Research carried out for practical goals
S	Organised Scepticism	E	Expert	Researchers are valued as expert problem-solvers rather than creative "organic intellectuals"

This led to the spread of various practices which had previously been extraneous or restricted to the fields of industrial and military research: there was now a reticence to make all acquired scientific knowledge available to the general public or alternatively concealed or explicit conflicts of interest.

The entry of market-based logic into academic science created a quadrilateral (science, politics, public opinion, economy) where there had previously been a triangle. This development forced science to rethink itself and its goals, and meant that politics, today more than ever before, was asked to perform its functions: to guide the development of society towards a desirable future.

Apart from the distortions created by market supremacy in certain fields of techno-science, one aspect deserves particular attention: the inequality in access to knowledge. With the criteria of the market economy, knowledge became a good just like other goods. The owner of knowledge tends to keep it rather than spread it. Those with greater economic resources have greater possibilities of becoming the "owners" of new scientific knowledge, so that unequal access to knowledge among nations and within nations produces the greatest inequalities.

Communication and Scientific Citizenship

As we have already mentioned, the construction of the society and economy of knowledge has been a lengthy process which began after the Second World War and which, although split into different phases, has constantly been informed by scientific research. Throughout this long process, during which the traditional relationships between the scientific community and the rest of society have altered, the public communication of science has gradually acquired both a new role – almost a new ontological statute – and new forms.

We should take account of this new role and these new forms if we wish to construct a democratic society based on knowledge, by examining at least four aspects of the changing relationships between science and society:

- *The ivory tower has crumbled.* The walls that have long divided the citadel of science from the rest of society have been broken down. The traditional separation has disappeared. We have entered a new era of organisation of scientific work which the English physicist John Ziman defined as a post-academic phase, distinguished by the fact that important decisions for the development of science are increasingly taken by scientific communities in participation with a series of other social groups. All this has "forced" scientists to establish an increasingly close-knit network of relations (and therefore of communication) with a public of non-experts.
- *Essential communication.* The era of knowledge is marked both by the invasion of society by science and the invasion of science by society. Society is increasingly well-informed about scientific knowledge and technology thanks to new scientific knowledge. Scientists are forced to communicate with a public of non-experts in order to take important collective decisions for the development of science in co-participation with them. Society, in its various forms, has an increasing need – an inescapable need – to be informed about science; this need has become decisive for citizens at every level, both individual and collective. In the era of knowledge and new relationships between science and society, the public communication of science (the communication of science to non-experts) is no longer a desirable but optional extra; instead, it is now an inescapable, twofold necessity: a professional necessity for scientists and a widespread essential need for the rest of society.
- *Complex communication.* The system of public communication of science consists of a large number of elements (various groups and social players) who have multiple non-linear relationships with each other. The flow of public communication of society takes place through a constellation of various important channels – some clearly visible, others more concealed – and defines a largely chaotic and unpredictable dynamic which is a typical aspect of complex systems.
- *Scientific citizenship.* The establishment of scientific citizenship is an essential element of a democratic society of knowledge. In the construction of the scientific community, the public communication of science is required to perform the decisive role of a lymphatic system.

In knowledge society, the development of relationships between science and society can evolve along two diverging directions. One is of an authoritarian type, based on the assumption that decisions on techno-scientific issues are too complex and need such rapid time-scales that they should be referred to specialists to avoid ethical and/or social chaos. The other one is of a participatory nature, based not only on the principle that even the decisions regarding the application of scientific knowledge should be taken on a democratic basis but also on the principle – dear to Francis Bacon, one of the pioneers of the new science – that new scientific knowledge should not be for the advantage of a few individuals or a few nations but for the whole of humanity.

In the first case, we are confronted by a model of authoritarian techno-scientific democracy, which in the ethical field can lead both to forms of absolutism (an ethic which imposes itself on others and becomes general law) and in the ecological field can lead to forms of militarisation of the local area (a techno-scientific choice imposed on citizens).

In the second case, that of participatory democracy in the era of knowledge, we face the need to construct *scientific citizenship*. This enables society to take decisions of a techno-scientific nature without falling into the trap of chaotic and paralysing populism. It also allows society to take advantage of all the opportunities provided by the development of knowledge as well as to minimise risks and socially undesirable effects (such as the increase in inequality caused by the difference in access to knowledge).

What actually does the concept of *scientific citizenship*, which gives a democratic form to knowledge society, involve? And how is it possible to construct it? There are no definitive answers to these questions. One sure thing is that *scientific citizenship* cannot be reduced merely to its *political* dimension, or, in other words, to a democratic method for making choices that involve science in ethically sensitive decisions.

This does not imply that the *political dimension* of the *society of knowledge* has a marginal role. If it is a question of deciding the location of a rubbish dump or the procedures for medically assisted procreation, the living will directive or strategies for combating climate change, it is necessary to find the best practices and the best forums where decisions can be taken without giving up the principle of maximum efficiency or the principle of maximum democracy. In other words, it involves finding the points of equilibrium where shareholders – i.e. the institutions of delegated democracy and experts – and stakeholders – those with a stake in the issues – can discuss and jointly participate in decision-making, each party with their own prerogatives.

It is far from simple. For this reason, a mature *scientific citizenship* also needs to develop, and as a matter of priority, in its *cultural dimension*. *Scientific citizenship* means exercising one's rights in an informed way. This involves the great themes of the public communication of science, beginning from the great centres of knowledge transmission: schools and mass media, given that the cultural dimension of scientific citizenship consists both of the maximum degree of rigour and effectiveness in communication but also, and above all, in the right of access both to information and the production of information. This is the crucial theme of the *cultural divide*, of the new inequalities within and between nations. Lastly, there is a problem of quality. Information and knowledge are not synonymous. Information is a clearly definable term which can be (and is) measured in quantitative terms. Knowledge is a highly sophisticated manipulation of information which requires a great capacity to create links between people and between disciplines. Scientific citizenship must therefore be structured according to its cultural dimension, the accuracy of information, access to information and knowledge and the quality of knowledge.

Only if the cultural dimension is full and rich, it is possible to develop another dimension of *scientific citizenship* in a satisfying way: *social citizenship*. Essentially, this implies not just having democratic access to information and knowledge but also a vast redistribution of its benefits. Not only science, but also the applications of science, must be for the advantage of the whole of humanity. In this sense, the environmental quality of development is a matter of crucial importance. An economy based on knowledge is only socially sustainable if it is ecologically sustainable. And vice versa.

We have therefore come to the fourth dimension *of scientific citizenship*, namely *economic citizenship*. Today there is a considerable tension between large companies in securing the monopoly of information and knowledge. This is a negative development because it is proven that scientific creativity is greatest where there is free circulation of information and knowledge. This is not just because it has been proved that the development of an economy of knowledge is greatest in a suitable environment with a strong propensity towards innovation; it is also due to the fact that it is possible to develop an economy of knowledge from the bottom-up – in other words, from small firms, especially cooperatives, which produce goods and services with a high degree of additional knowledge. Only by developing the *economic dimension* of *scientific citizenship*, i.e. by combating monopolies and oligopolies of knowledge and extending the opportunity of producing goods and services with a high degree of additional knowledge within society, can we actually begin to consider solving the greatest contemporary social problem: inequality.

In conclusion, we live in a world which is increasingly characterised by scientific knowledge and technological innovation. We therefore need to extend scientific citizenship. We still do not know how to define the concept of scientific citizenship in sufficient detail. However, we do know that every level – whether cultural, social, political or economic – implies participation: the extension of democracy, both in formal and substantial terms.

It also implies communication: public and transparent communication. As in the seventeenth century, science is once again being called on to break down the "paradigm of secrecy".

Science Communication at Glance

Luigi Amodio

Introduction

Recalling his childhood in his autobiography, Albert Einstein mentions this touching episode:

> A wonder of this kind I experienced as a child of 4 of 5 years when my father showed me a compass. That this needle behaved in such a determined way did not at all fit into the kind of occurrences that could find a place in the unconscious world of concepts [. . .]. I can still remember – or at least believe I can remember – that this experience made a deep and lasting impression on me.

Naturally, it is impossible to say whether this experience was really the "moment" that gave the impulse to one of the most extraordinary adventures of contemporary science, indeed the most extraordinary. What is certain is that the words of the great German physicist enlighten us, partly due to their passion, about the importance that is played – in the incessant dialogue between the subject of knowledge and the world – by wonder linked to the experience of reality. Indeed, Einstein says that his predominant emotion was that of wonder, astonishment and enchantment which did not derive from an astonishing or "monstrous" event. It is a small object, a needle, with its unexpected behaviour, that reactivates the chain of emotions and adult reflections. And, it should be emphasised, it is a real event, an actual phenomenon, rather than a simply conceivable event. The reality of nature and its "simple" phenomena left a mark on the young Einstein that was to be profound and indelible.

I believe that these two dimensions, wonder and direct contact with phenomena, are the ones that present the biggest challenge to those currently involved in popularising science and scientific method. It is a challenge which, in my opinion, has been met successfully if one considers the wealth of experiences in a sector which – in the world of research and the media – seems to be expanding on an international scale.

L. Amodio (✉)
General Director of Fondazione Idis-Città della Scienza, Naples, Italy
e-mail: amodio@cittadellascienza.it

A.-M. Bruyas and M. Riccio (eds.), *Science Centres and Science Events*,
DOI 10.1007/978-88-470-2556-1_4,

Communication in Science Museums

The aim of this chapter is to offer an interpretation of the scientific communication methods currently employed in contemporary museums and new generation science museums. A particular attention will be paid also to the changes taking place within them such as for example the general crossovers between cultural and economic practices, local transformations and the introduction of new communication technologies. A large part of the ideas and contents presented here derive from the observation of structures involved in the practical issues of communication procedures and particularly in contexts that refer to the transformation of the metropolis and the emergence of new skills and new professional profiles. Most of the observation is obviously linked to the activities carried out in Città della Scienza, the science centre in Bagnoli, near Naples, where the current author has worked for many years, as well as in a range of European museums and science centres, contemporary art museums and places linked to the tourism industry. The following themes will be discussed: the distinction between science museums and new science centres, with particular reference to aspects of their transformation; the communication practices that take place there, working with the various media, and the techniques used; the concept of virtual reality and the role, including the instrumental role, of new technologies; lastly, there are a few reflections on the relationship among science, society and citizens.

From Museums to Science Centres

We can begin from the definition of "museum" taken from the Oxford Paperback dictionary and try to analyse it:

> [Museum] *noun* Building for storing and exhibiting objects of historical, scientific, or cultural interest. museum piece 1 object fit for museum. 2 *derogatory* old-fashioned person etc.

Three concepts immediately stand out. The museum is a place for storing and displaying objects. These objects, documents and finds are generally old, rare and *therefore* valuable and important. Lastly, the word is associated with uselessness and age: "museum piece" is almost an insult, even though intended to be humorous.

The inversion of these three concepts – the subject of the following discussion – represents the heart of this book. The places analysed here do not conserve objects, but display experiences; the actual objects they contain are not rare and only in a few cases do they have value *in their own* right, but the experiences they offer visitors are unique; lastly, since they are "mutants" and constantly changing, they adapt to the contemporary world and its needs.

But what came before the "museum", in the sense of the modern western institution with which we are all familiar and which we instinctively associate with this name?

In the past, until the Middle Ages, there was undoubtedly the need to display, especially as a symbol of power: precious objects, booty, liturgical objects, exotic natural curios and so on. Stretching the meaning of the word, even *potlatch* could be included as ancestors of modern museums, since they derive from the same insuppressible need to display power by parading objects. The great museums – the Louvre and the British Museum, to mention just two famous examples – constitute real imperial warehouses, living catalogues of the power of the empire, flaunted *showily* as a warning to defeated nations and military adversaries.

With the advent of Humanism, the rational need to order collections prevailed. The *studioli* (cabinets) of Lionello d'Este at Ferrara, Federico di Montefeltro at Urbino, Isabella d'Este at Mantua, etc. are interesting examples. The *studiolo* of Francesco I de' Medici at Florence led to the foundation of the Uffizi Gallery, the first modern museum, in which the dissemination of culture was for "all", and a response to the emergence of new social classes and a new social organisation.

Later on, the *Wunderkammern* in Northern Europe and the *Cabinet de Curiosité* in France became the precursors of modern science museums which resembled – as we shall see – a form of pre-academic science, often cultivated in private.

With regard to science museums, it is important to make a preliminary distinction between different "generations of museum".

First generation museums can be defined as museums based on the principle of "look but don't touch!" where objects and collections of (often rare) finds are generally kept and protected, similar to the principle underlying art, archaeological and historical museums. In some of these – and, specifically, those devoted to technology – the first interactive experiences appear at the beginning of the twentieth century, based on the principle of "pushing the button", in order to activate the reproduction of scale simulations of machinery, workings and other mechanical devices. As it is obvious, the dominant philosophy in first generation science museums is quite clearly positivist. Often derivations from the Great Universal Exhibitions – and often placed in the pavilions made to house the exhibitions at the conclusion of the events – these museums are a celebration, to quote the Italian poet Leopardi, of "magnificent and progressive destinies"; if we consider the case of Munich, the museum explicitly declares its "educational" nature intended for the working classes and citizens of industrial metropolises. While the Natural History Museum in London (founded in 1881) represents a significant example for the natural sciences, other institutions such as the Deutsches Museum in Munich (1906), the Science Museum in London (1928), and the more innovative and "radical" Palais de la Découverte in Paris (1937) fully represent this type of museum. Exhibits of the first generation therefore include objects, finds, machinery of various types, characterised by inertia and untouchability, protected by glass display cases and screens.

The second generation of science museums corresponds to those which we shall refer to henceforth as Science Centres, a term which, even in the Italian experience of the *Immaginario Scientifico* in Trieste and *Città della Scienza* in Naples, is not translated in most cases. Set up at the end of the 1960s with the creation of the Exploratorium in San Francisco and the Ontario Science Centre in

Canada, these institutions completely overturned the approach of traditional museums, paradoxically reviving – as we shall see shortly – the tradition of displaying science "in action". These "museums" are not collections of inert objects but "hands-on exhibits": essentially, they display *experiences* whose value does not lie in "exchange", so typical of objects displayed in traditional museums, but in "use". This use value derives from emotion, wonder, and the experimental content that these exhibits manage to convey. The watchword is therefore "forbidden not to touch!" and the philosophy of science centres can be summed up by the inscription at the Exploratorium, "If I listen I forget, if I see I remember, if I do I understand", which succinctly conveys the communicative essence of interactivity. In this context, we have also witnessed the introduction of personal computers and multimedia to exhibitions, a significant change which would pave the way for later more radical innovations in style and content. At a historical level, it is worth underlining the success that this model of scientific communication soon began to enjoy worldwide. In Europe the first science centre to capture the attention of the general public and scholars of communication – and not just scientific communication – was the Cité des Sciences et de l'Industrie, known more informally as "la Villette", by the name of the Paris quarter where it was established in 1986; it also represented an international model of reuse of an area that had previously been productive (slaughterhouses and markets) and which has now been given a cultural function.

It is difficult to describe a hands-on exhibit. We could just say that a typical example of this kind of exhibit reproduces a natural phenomenon – preferably a physical one – together with captions made up of three sentences: "What to do", namely how to start the device, "What to observe", in other words, what changes, the behaviour of objects, measurable and observable variations of parameters to look out for and "What happens", the scientific explanation in everyday language of the law that lies behind the observed phenomenon. Along with these devices, another distinctive hallmark of science centres is the reconstruction of "laboratories". This involves simplified laboratory activities which consist of didactic activities and modules and demonstrations, small scientific displays conducted by explainers in cooperation with, and with the participation of, the public.

But who can be claimed to be the inventor of modern science centres?

Born in New York in 1912 into a family of German-Jewish origin, Frank Oppenheimer – the younger brother of Robert, an important physicist and one of the main members of the Manhattan Project – developed a keen interest in science from early childhood, as did his brother. A student of physics at Johns Hopkins University, he graduated in 1933, later moving to England to study at Cambridge University. He completed his studies there before returning to the United States in 1935.

During this period he met and later married Jackie Quann – an economics student – whose role in the life of Oppenheimer and the creation of the Exploratorium was of great importance. In 1937, they joined the Communist Party together; like many young intellectuals of the time, they were struck by the rise of Fascism in Germany and Italy and by the revolution in Spain. They both left the Communist Party in 1940 but the experience (as we shall see below) was to be a decisive factor in the choices they made later on in their life, and indirectly influenced their decision to set up the museum in San Francisco.

As we know, between late 1938 and early 1939, the military implications of the research conducted in Germany into nuclear energy spread throughout the world and soon even scientists who until then had refused to carry out research in the military field felt they no longer had any other choice. The story of the Manhattan Project is well-known: on 16 July 1945, the two Oppenheimer brothers had the chance of seeing, from a protected bunker, the first nuclear explosion in history and a light "brighter than a thousand suns", as it was later described by Robert. After the end of the Second Word War, Frank Oppenheimer returned to work in California on the development of the first linear proton accelerator. He later accepted a post at the University of Minnesota to carry out research on nature and the origin of cosmic rays. Meanwhile, however, the American political situation had undergone profound changes and the USSR had become the main enemy. The Cold War had reached its height.

Oppenheimer and his wife were soon "eliminated". Frank Oppenheimer had to abandon his university career and, together with his family, he retired to a ranch in Colorado. They thought they would only spend a brief period there but they ended up staying for 10 years. It was during this unusual state of exile that Frank and Jackie gained the experiences that led to the creation of the Exploratorium. After a period spent working the land, Oppenheimer began to teach science, biology, chemistry and physics at the school of Pagosa Springs, where he ran a course on electricity in the school (with only one classroom!) which was also attended by his own children.

It was this experience that led to the development – or, rather, rediscovery – of the approach to the study of science based on direct experience which became the trademark of the Exploratorium. The pupils were asked to use the surrounding environment as a research area for mechanical objects or natural samples to use for their experiments. News about this experience soon spread and in 1959 Frank accepted an offer to take part in a training course for science teachers.

Meanwhile, the launch of Sputnik by the USSR (1957) had sparked off a passionate interest in the study of science in America which almost became a patriotic activity. As a consequence, science curricula were completely revolutionised; in particular, it is worth recalling the Physical Science Study Curriculum, developed by Gerald Zacharias at MIT, an old friend of Oppenheimer, who was soon involved in the development work. From the PSSC came the idea of Elementary Science Study, which finally convinced Oppenheimer and his fellow workers of the extraordinary power of direct interaction between children and scientific phenomena and the potential of this teaching method for all age ranges and educational levels.

After being readmitted into the university world, Oppenheimer noticed that not much had changed in the way science was taught and in the approach of students towards scientific disciplines, which appeared to be instrumental and lacking in intellectual enthusiasm. Science had principally become an opportunity for remunerative careers on an economic, rather than intellectual, level. At the same time, the huge developments in science and technology had gradually distanced these fields from daily life, so that people's attitudes tended towards blind faith, or possibly fear, but not the desire to understand and master. In this context, Oppenheimer's attention had decisively switched from research to teaching and his work during this period enormously contributed to reinforcing his role in this field. This led him to receive a grant from the National Science Foundation to plan a new model of teaching for the introductory physics course; thanks to this funding, he set up a "Library of Experiments" (one hundred) which represented the prototype for the Exploratorium.

In 1965 another important and decisive episode took place: during a research trip to Europe, Oppenheimer had the chance to visit and study three of the most important science museums of the time: the Science Museum in London, the Palais de la Découverte in Paris and the Deutsches Museum in Munich. Analysing the three institutions, he came up with the idea of setting up a large science museum in America, which would be marked by some significant innovations, most importantly in terms of contents and communicative approach. In particular, attention would be focused on knowledge processes rather than disciplines and a conscious attempt would be made to avoid creating the atmosphere of exalting technique, typical of the museums of London and Munich, which had been designed as places for educating citizens (or rather, citizen-workers) about the extraordinary power of industrial development. Instead, the interactivity between the visitor and the exhibits would be extolled to the maximum, overcoming the fear of devices which could be activated simply by pushing a button (also extremely innovative, compared to traditional museums where the dominant approach was "to look but not touch" collections of finds protected behind display cases).

In this atmosphere, Oppenheimer formulated the metaphor of a "walk in the woods" to describe his idea of museum: "A place where people could come, alone or in a group, to investigate the world of nature and share their own discoveries with others. But what we see in the woods is restricted by our senses, helped perhaps by a magnifying glass. In the museum, special tools extend the field of human observation to reveal a world which is normally concealed from us".

From that moment on, his decision to create a museum became operative and the practical work began; this was to involve Jackie Oppenheimer directly (she was responsible for choosing San Francisco as the site for the museum). Between 1967, the year when the first surveys were made in the Californian city, and 1969, the year when the Exploratorium was opened, the Oppenheimer's work was largely of a "political" nature. It was necessary to create a consensus and, above all, to raise funding. However, it was also a phase of hugely creative work, beginning from the name of the museum: the first idea was MOSAIC (Museum of Science, Art, Industry and Crafts), but the structure gradually became referred to as an "Exploratorium" and the Oppenheimers chose to opt for this name. In February 1969, it was decided that the site of the new museum should be the former Palace of Fine Arts, a large structure built in 1915 to celebrate the opening of the Panama Canal and situated in the charming district of Marina.

Faithful to their philosophy of interactivity, the founders of the Exploratorium decided to open the museum before creating its contents which would be made in front of the public and indeed, actually encouraging the public participation in the creative process. On August 20, 1969, this new strange museum was opened. The visitors arrived almost immediately. Tourists and curious onlookers who were enjoying themselves in the park discovered the museum mainly by accident and entered to see what was going on. There was almost nothing on display, just a small group of people hard at work and a sign saying "Here is Being Created the Exploratorium, a Community Museum Dedicated to Awareness".

Since that day, the Exploratorium has become an essential point of reference for everyone involved in scientific museology. Apart from judgements or individual tastes, this museum represented a real revolution in understanding the relationship between the sources and targets of communication, even more than giving shape to the contents on the ground.

After this lengthy digression, devoted to the first *science centre*, we come finally to the third generation science museums. In this case, the characteristics are essentially to be found in the role of the new information and communication technologies which constitute the main means of interaction. Another key element concerns the informative aspect, which prevails over the traditional "educational" and "training" aspects that were typical of previous generations. Lastly, the experience of visitors operates at various dimensional levels, since new technologies break down the space-time barriers that marked the ways of using the museum in the past (without, as we shall see below, dissolving – indeed, quite the opposite – the "physical" nature of the "museum as institution").

An example of a third generation exhibit is the TELEZONE of the Ars Electronica Center in Linz. A robotic installation with Internet components enables the Web Community to plan and build architectural structures and share information with a virtual community of other users. It represents the architectural evolution of Telegarden, a similar installation which enabled remote users to "tend" a flowerbed placed at the entrance to the centre.

The Austrian museum perfectly represents the model of a third generation structure: new technologies constitute the main, if not the only, intelligent interface between the visitor and the museum, conditioning the actual physical structure of the building. For example, through the use of sensors, the monitors dotted around the structure are adapted to the height of the visitor; besides having its own function, the lift is an artistic installation among others. The virtual presence of online visitors described above breaks down the barriers between the building and the outside world, extending its installations into time and space.

Another more traditional example is the Wellcome Wing of the Science Museum in London, an interactive area based extensively on new technologies, and completing the traditional visit to the museum. In this context, new technologies feature mainly as an extension of the museum's contents beyond its boundaries and a permanent tool for creating dialogue with visitors; an example is the web site which makes it possible to "personalise" some of the functions and place some of the experiences of the exhibition area into a "virtual briefcase".

The Wellcome Wing is also one of the best examples of the success and diffusion of the model of interactive scientific communication in museums, representing a striking case of successful innovation of a traditional museum according to the "science centre typology" (an approach also being pursued in historic institutions like the Deutsches Museum in Munich, the Palais de la Découverte in Paris and many natural history museums).

This success is also reflected by the increasing frequency with which new science centres are opening throughout the world and the creation of networks that now exist worldwide: the ASTC in the United States, ECSITE in Europe (with its English and German branches), the Indian National Council of Science Museums, the network operating in the Southern East Asia, and so on. This situation has led the various networks to set up a world event for members to meet and exchange experiences and best practices at the World Congress of Science Centres which takes place every 3 years. These events provide the opportunity for dialogue among distant places (the world congresses took place in Helsinki in 1996, in Calcutta in 1999, in Canberra in 2002, in Rio de Janeiro in 2005, in Toronto in 2008, in Cape Town in 2011).

Some Further Considerations

This tripartite scheme is obviously not the only possible way of "organising" the wealth and variety of experiences included under the label of science centre. One of the main contributions to the definition of typologies and models of science centre is the study carried out by the Fondazione Giovanni Agnelli entitled *L'esperienza internazionale degli Science Centres*. Although it dates back to 1998, it remains one of the clearest analyses in Italian of these new structures of scientific communication. The study analyses 12 cases in Europe and the United States, taking into account – for each of them – a series of indicators: planning, cultural basis, logistics, services and facilities, resources, and prospects for development. The study led to the definition of 6 typologies which are briefly described below.

The *first typology*, summarised as Educational service centres: teaching science and technology, refers to the Exploratorium in San Francisco, an institution which has already been discussed above and whose importance does not need restating. Here it is worth emphasising the recent "shift" in its role whereby the Exploratorium provides support for the school system through participation in activities – sometimes in association with other institutions – in the field of building "bridges" between formal and informal learning.

The *second typology* – a completely different one – is that of the Scientific fairgrounds: having fun with science and technology. Although based on collections of hands-on exhibits, the main aim of these centres, such as the Pacific Science Center in Seattle mentioned in the study, is to surprise and entertain visitors: they are set out to offer an alternative to theme parks and funfairs. Less sophisticated from a scientific and educational point of view and clearly more focused on marketing and promotion, scientific fairgrounds contribute to the dissemination of scientific knowledge, albeit in a very simplified (and perhaps rather trivialised) way.

In the *third typology*, the authors of the study include those institutions that have devoted particular attention to the links between science and technology and their social effects. Scientific agoras: discussing science and technology is the term created to define them; two examples are included in this category: Nemo (formerly new Metropolis) in Amsterdam and the Wellcome Wing of the Science Museum in London. As we shall see, both these institutions have now been "surpassed" by new bodies currently operating in Europe but when the study was carried out, they represented groundbreaking ventures on the international scene (indeed, at the time, the Wellcome Wing was still at the planning stage).

Exploring the frontiers of science, particularly its technological applications, and presenting them to the public are the functions of the so-called "Innovation Monitors", which the authors include as the *fourth typology*. In this case, all the mentioned examples cited are located in the United States (whereas similar European institutions are more oriented towards scientific education as well as the presentation of technological applications): the Columbus Center in Baltimore, the Inventure Place in Akron and, in particular, the Tech Museum of Innovation in San José, the authentic "museum" of the Silicon Valley. It was sprang up as an offshoot for research and development in the field of *computer sciences* and it contributed enormously to the economic and social wellbeing of the area marking, from a cultural perspective, such an important chapter in human history.

The penultimate type is that of the Science Cities, institutions designed to offer universal scientific communication or, to use a fashionable term, "generalist" institutions whose aim is to provide methods and tools for orienting oneself in a world increasingly filled with science and technology. The example given in the study is the *Cité des Sciences et de l'Industrie in Paris*, better known as La Villette by the name of the district where it is situated. This area has undergone a process of urban upgrading that reflects the intentions of the French government which was highly committed to the project and still contributes enormously to its success and maintenance. La Villette embodies the undertaking to build a national centre of excellence whose effects – in terms of its activities and overall impact – concern not only the capital but the whole country in all fields of scientific knowledge and technology.

The last typology regards "incubators of local development" where science and technology mainly constitute factors of economic growth. The term refers to institutions which, besides the methods and tools they employ, are mainly aimed at revitalising a local area through the promotion of scientific information and the culture of innovation – factors considered as essential ingredients for the economy of the future, based primarily on knowledge. Città della Scienza in Naples combines in one location the educational and demonstrative functions typical of a science centre and those of Science and Technology Parks (Business Innovation Centre and Centre for Advanced Training). It is the institution chosen as an example of the last typology and also represents the only Italian institution considered in the study.

The Current Situation

This brief overview gives us a highly dynamic impression of the state of science museums, a "family" which is growing at a striking rate and whose force lies in the diversified experiences, styles, roles and forms of interaction. Moreover, the debate that used to pit traditional structures against innovative ones is dying down and is being replaced – at least in the more advanced points at the level of theoretical formulation and practice – by a mutual attempt to overcome the distinction between "museums" and "science centres"; at the same time, display and exhibition methods (see the case of Cosmocaixa, previously Museu de la Ciencia in Barcelona) are becoming increasingly mixed and the key word seems to be continuous change and development towards new configurations. Meanwhile there is an increasingly skilled community of professionals which is growing, both in quantitative and qualitative terms; this community is becoming more organised and setting up new centres for formulation and joint action.

The Evolving Relationship with the Public

One important aspect worth examining is the relationship with the public and the more general role of cultural institutions of this kind.

Firstly, science centres tend to place the needs of the public at the centre of their own activities, as opposed to traditional museums which concentrate on conservation. This emerges very clearly not just at the level of planning and museography but also from the consideration of the public as "users" in flesh and blood of a physical space. This has led the attention to be focused on "public services", ranging from refreshments to access for the disabled, to rest areas to break up the visit.

Secondly, as well as the traditional functions of conservation and enhancement, new functions are becoming better established. Museums and centres are thus becoming a "services agency" for public clients and/or end users, partly with the purpose of meeting objectives of an economic nature. Besides providing traditional guided tours, it is increasingly common to find teaching and educational activities for schools and the public, activities for small children, refresher courses for teachers, technical assistance for schools and other bodies in innovative educational projects, integrated communication projects for companies and public institutions, careers guidance services, support with planning and designing exhibitions and exhibits for other museums, organisations, companies etc.

Also in terms of contents, we are witnessing important transformation processes. Increasingly, institutions are being set up where it is difficult to find completely defined thematisation; this is particularly the case in the relationship between art and technology. Examples include the Ars Electronica Center in Linz, Austria or the *Zentrum fur Kunst und Medien* in Karlsruhe, Germany which "hybridise" art, science and technology in structures which are also open to the typical public of a science centre, such as children. On the other hand, *besides contents*, another evident phenomenon is the coordination of tourist marketing into "chains" of museum-logos. They are mainly distinguished by the link with the local area such as projects of urban upgrading and the relaunch of local marketing, although disconnected from the local area for historical reasons, and legitimised by the logo as a "brand of quality". As an example, we can mention the Guggenheim museums and, above all, the proverbial case of Bilbao or the more recent Tate Modern in London.

This overview of the state of the art in the field of science museums can lead us, in conclusion, to the formulation of a diachronic table – related to the history of the western world – which summarises the potential links among four parameters:

- Economic and local organisation;
- Structure of society;
- System of scientific research;
- Type of museum/display.

As one looks at this table, particular attention should be paid to what happens in the final phase. While the history of the evolution of museums in terms of the organisation of "academic science" represents a heritage of consolidated knowledge, the history of the last few decades is quite a different story. The spread of mass communication and increasingly pervasive new technologies, which also affect the organisation of work, is closely intertwined with a new existential dimension in which the continuous updating and permanent adjustment of knowledge transform what used to be simply users into producers of knowledge and information where the knowledge economy puts consumers "to work", enhancing the activity of cultural consumption. Second and third generation science museums, with their strong emphasis on interactivity and visitor participation, constitute a highly representative and symbolic example of this trend. This phenomenon is accentuated in institutions which mainly focus on new technologies where even interactive objects are replaced by products based uniquely on information (multimedia, simulations, multimedia interfaces, etc.).

In conclusion, we can state that today's science "museums" appear to be becoming increasingly less conservative, increasingly detached from the history of the local area, increasingly rooted in architectural works or local marketing plans, and increasingly run as companies; meanwhile, however, they are full of additional services (reflecting their new nature) for visitors and the general public. Lastly, they are increasingly sensitive – for better or for worse – to visitors' needs.

Museums, Science and Society

As we have already seen, the Exploratorium represents a crucial turning point in the history of scientific museology. Before discussing the main theme of this chapter, it is worth briefly recalling the salient features of the Exploratorium, whose model has inspired and led to the establishment of dozens of similar institutions throughout the world.

The first distinctive feature is interactivity, based on the display of experiences rather than objects endowed with value in their own right, like those traditionally conserved in museums. *Hands-on exhibits* – objects which need to be handled to generate the reproduction of phenomena – represent the creative evolution of similar devices already present in large European science museums (e.g. the Deutsches Museum in Munich or the Science Museum in London); for several decades, they have become the main point of attraction of *science centres* worldwide. Besides the trivialisation often created by the minimal usage of the potential of hands-on exhibits, interactivity naturally refers to the intrinsic characteristics of science, in other words its experimental nature.

The second feature that clearly emerges from the first *mission statement* of the Exploratorium, subsequently simplified and updated, is the democratic vocation of science and its socialisation. "The Exploratorium's mission is to create a culture of learning through innovative environments, programs and tools that help people (of all ages, origins and backgrounds) nurture their curiosity about the world around them".

The museum, or rather the *science centre*, is not just a space open to the participation of everyone so that science and its effects are shared as much as possible. It is also a training ground based on equality ("...people of all ages, origins and backgrounds..."), which operates first and foremost both on the cognitive level ("...exploiting one's own innate curiosity...") and on the educational level in a society, like that of the United States, which is an ethnic and cultural melting pot.

The third and final aspect is the aesthetic dimension. The Exploratorium – originally founded as a "museum of art, science and human perception" – quite explicitly seeks to create a relationship with its visitors. It uses the beauty and implicit elegance of natural phenomena which represent a sort of "bait" to capture attention of visitors and create a sense of wonder that can then be channelled towards the understanding the laws involved. Beauty returns, this time in an explicit form, in the decision of the exhibition curators of the Exploratorium (and other museums and science centres) to entrust artists with the creation of hands-on exhibits.

It was on these foundations – extremely summarised briefly here – that a whole generation of museums arose or traditional museums were modernised with undisputed success, at least until the mid-1990s. At this point the transformations linked to the role of science and technology in contemporary society became explicit and forced those working in this field to come up with new ideas.

The first transformation regarded the epistemological and organisational structure of scientific and technological research which can be summarised as the transition of academic science to its new post-academic condition.

The term 'academic science' is used here to define what we usually mean when referring to 'pure science' or science in general. Its characteristics emerged in western Europe during the scientific revolution of the seventeenth century and its procedures, formulated by Robert Merton, are well-known: communalism, universalism, disinterestedness and humility, originality and scepticism.

The advent of post-academic science emerged in the period following the end of the Second World War and has only recently become evident; it depends both on factors that are external to science as well as on internal factors and therefore on increasingly rapid scientific and technological progress and the increasing interdependence between science and technology.

As John Ziman wrote, the characteristics of this new condition of science are the following: collectivisation, limits to the development of science, the enhancement of knowledge, the politicisation of science, industrialisation and bureaucratisation.

However, the most interesting aspect in this context is that the number of those involved in scientific work in the post-academic framework is growing continuously; it can now be said that scientific communication for non-experts has become an activity which is completely internal to "doing" science, and an important activity for its own development.

According to Pietro Greco:

> This new era in scientists' method of working has led to a redefinition of the role that scientific communication for a public of non-experts could have for the development of science itself, as well as for the cultural and civic growth of society in general. The hypothesis is therefore that the public communication of science plays an important role in the development of science itself.

Secondly, the nature of contemporary science and the change in paradigm caused by the advent of the new life sciences firmly repositions the theme of the impact of science *on* daily life and *on* society. The current perception of science is increasingly linked to its possibilities of reaching the roots of existence itself, through modern biotechnologies, nanotechnologies etc., thus opening up unprecedented problems of social, political, legal and philosophical nature. In this context, there are clear limitations and deficiencies in either a technocratic approach, whereby "experts" are the only ones legitimised to express their opinion, or a bioethical approach, which refers solely to the moral values of the individual. On the other hand, according to the sociologist of science Massimiano Bucchi:

> It is important to understand that the participation of non-experts in technical and scientific processes and the participation of scientific experts in public debate are largely two sides of the same coin; they have common roots and, indeed, tend to feed off each other.

There is an increasing need to create new forms of dialogue and discussion among science, society and citizens which are much better organised than anything that has yet happened.

The last aspect to consider can be defined as a "linguistic turning-point"; it is linked to the computer revolution whose initial conditions and consequences could be broadly foreseen as early as the 1960s and 1970s. In particular, the increasingly widespread use of new information and communication technologies to produce both material goods, in automated factories, and immaterial, and relational, goods and services has given back "language" – and, more generally, the manipulation of symbols – a central role. One of the main consequences is the subsumption – in the production of value – of work carried out unwittingly by the end users of services, who thus become wealth-creators even in those spheres of life that used to be considered "free". This aspect of contemporary capitalism – and of the system of organisation of the mass media – represents one of the main peculiarities linked to the increasingly large-scale use of new technologies in everyday life. This situation has been eloquently described by the theoretician of information society Manuel Castells:

> Processes of cultural transformation summarised under the ideal type of the network society go beyond the sphere of social and technical relationships of production: they deeply affect culture and power as well. Cultural expressions are abstracted from history and geography, and become predominantly mediated by electronic communication networks that interact with the audience and by the audience in a diversity of codes and values, ultimately subsumed in a digitalized audiovisual hypertext.

All this is also having clear effects on communication practices in science museums. It could be argued that, during the age of academic science, the adaptation of science into museum form took place essentially in the large science and natural history museums and subsequently in science centres; however, in the age of post-academic science and the widespread use of new information and communication technologies, we are witnessing the rise of new practices, both in terms of display and the use of museum settings, in these institutions which obviously underwent significant alterations. We shall look at several cases in Italy and in international contexts.

The first example is the Cosmocaixa in Barcelona, opened in September 2004, which represents the extensive redevelopment of the old Museu de la Ciencia, part of the Fondación Caixa de Cataluña. Jorge Wagensberg, director of the Cosmocaixa and now head of science for the entire bank foundation, has defined this experience using the concept of the "total museum". In brief, it consists in incorporating the features of various types of scientific exhibits (objects and finds, hands-on exhibits, living animal and plant organisms etc.) within a single institution; it also consists in reaffirming the trend towards unifying forms of knowledge which tends to be lost in modern science because of its gradual segmentation and specialisation. In this sense, Wagensberg summarises the content of his museum with the slogan "from Quarks to Shakespeare", emphasising that interactivity does not just consist of the hands-on approach of interactive exhibits; it also involves a "hearts-on" approach based on real living objects as well as a "mind-on" approach deriving from mainly abstract experiences.

The second case, the one of the Dana Centre in the Science Museum in London, reflects a radical transformation in museums where the real object of interest is dialogue. The Dana Centre was opened in 2003, a few years after the opening of the Wellcome Wing, which had already introduced a hands-on approach to the historic London museum, based on new technologies and themes at the frontier of science. The Dana Centre bases its activities on encounters with experts linked to the "hottest" themes in contemporary science. It represents a sort of extension-evolution of the Wellcome Wing. Graham Farmelo, who was the first director of the Wellcome Wing, states that:

> The new science centres, the best type, guide reflection on contemporary themes, looking forwards; they stimulate public interest in the future and prompt the visitor to wonder 'what might happen'. Another important feature is the capacity not so much to talk to the public but rather to have a dialogue with it in order to create an opportunity for a response, so that the public can reply "we don't like this". This is very important and extremely different from the situation 10 years ago.

The Dana Centre has an informal atmosphere and is "prolonged" in time and space through its web site, making it possible to take part in forums, chats and in-depth analysis of the themes dealt within the centre. The centre revives a deep-rooted English tradition, that of spending time in the pub; however, in this case, the subject is science and technology and their frontiers. It can really be said that in this case it is the visitor who, if suitably stimulated, constructs and performs the activities that take place. While in the tradition of scientific museology the "work" requested to the visitor is essentially of a "physical" nature (turning on the exhibits using simple commands), within the dialogic context of the Dana Centre the activities are primarily of a relational nature, based on language and symbolic exchange. The synthesis between production and consumption of the contents takes place at a higher and deeper level.

This is also the case for numerous projects funded by the European Commission within the framework of the Programme "Science and Society" of the Sixth Framework Programme for Research and Development. The Action Plan on "Science Society and Citizens", launched by the Commission in 2001, traces the guidelines and the theoretical motivations that have led to the decision to allocate a share of the budget for R&D to this type of action. This decision was to continue in the Seventh Framework Programme in which, however, greater emphasis is placed on the role of science, starting with the name of the programme itself which was changed to "Science in society". The programme consists of action research projects involving several institutions belonging to the "community" of science centres and science museums (from ECSITE, the European association of science centres, to Fondazione IDIS-Città della Scienza in Naples from the Cité des Sciences et de l'Industrie in Paris to the Deutsches Hygiene Museum in Dresden). The projects are based on the use, adaptation or formulation of participatory forms of discussion which often choose the museum context as the setting for organising participatory activities. This is not just due to the neutrality of the setting, but particularly to the opportunity to offer to the participating public a heritage of "resources" (of a material, exhibit, human and informative nature) which can fill the gap in knowledge, so often revealed by surveys and studies regarding the themes of frontier science and contemporary research, and which the media find increasingly difficult to fill. In this sense, this type of activity seems to move in the direction of museum communication that encourages in-depth analysis rather than speed of use, often uncritically considered to be a worthy objective in observance of the competition with the rapidity of information spread by old and new forms of media. Moreover, as George Gaskell of the London School of Economics has observed:

> It is remarkable that the radical proposals put forward 4 years ago in the Science and Society Action Plan are now widely accepted in Europe and America... Public values are changing, and in this context maybe we should be talking not about communication but about dialogue and engagement.

In conclusion, it can be said that there is an increasingly widespread tendency to re-evaluate the museum/science centre as a "forum", a meeting place and an arena for exchanging ideas. This can take place in fairly structured ways, linked to an extent to exhibitions, but which is also completely independent of them (as in the case of the "science cafés" organised by various museums outside their own premises or some of the projects mentioned here). This conclusion offers scope for reflections which could be the subject of further studies and future researches, linked to the message of the Exploratorium, which was particularly powerful in the period of its foundation in the political and cultural climate of the 1960s, but still active both there and elsewhere; the message regards the deinstitutionalisation of science and its communication to the public, the liberation of its potential in society, in order to give citizens the possibility of exerting some control, even in such a complex field dominated by "experts". Although many things are changing in museology and in the practices of museum communication, the strong cultural and political reasons that led to Frank Oppenheimer's decision to create the famous science centre in San Francisco continue to thrive in original ways, providing indications of great interest for the future.

Communication Tools in Museums

In this chapter we will attempt to provide a brief reply to the basic question "What should be communicated and how?". Therefore we shall begin by examining the communication of contents through the two basic elements of modern museum communication: exhibitions and events. We shall then provide a brief overview of communication aimed at constructing the target market before finally looking at social communication and local communication, using as examples campaigns and other activities linked to the social role of museums and science centres.

Exhibitions are the distinctive feature of museums while objects are the distinctive feature of exhibitions. Without objects (finds, works of art, interactive exhibits, etc.), exhibitions would be impossible. This seemingly banal concept is actually extremely important because it immediately places our reflection within a real space, the physical space of museum's rooms where the exhibitions are held.

We can classify exhibitions according to two variables.

The first variable regards contents. Typically, an exhibition features a collection of objects (e.g. a collection of ancient Greek vases or a collection of beetles from the Amazon rainforest etc.). It may show objects that exemplify a phenomenon or artistic, historical or scientific process (e.g. French Impressionism, evolutionism, nineteenth century chemistry etc.); lastly, through the display of objects, it may illustrate themes which are, in some sense, “abstract” (e.g. time, beauty, immateriality etc.).

The second variable regards length. In this case, there are permanent exhibitions which typically correspond to a museum’s collection, temporary exhibitions which have a clearly defined beginning and travelling exhibitions; the latter are a special type of temporary exhibition designed to be moved from one display site to another (and which can, during the movement between different premises, vary in terms of structure and the objects on display [e.g. for reasons of safety of the works themselves]).

An exhibition always involves displaying objects. If the theme of the exhibition cannot be presented using objects, there are many alternative means of communication which are considerably cheaper as well as easy to make. Examples might include a publication, a web site, an audiovisual feature and so on. Simultaneously, when an exhibition is being prepared, it is necessary to pay considerable attention to the quantity and quality of the exhibits.

Too many exhibits may prove excessive for the visitors’ attention span; too few objects may disappoint them and leave with a sense of dissatisfaction.

If the exhibits are all of extremely high quality, there is a risk that they will level each other out. It is therefore advisable to strike a balance – which should also be qualitative – in the exhibits presented to the public.

Since they are designed to illustrate ideas, arguments and cultural positions, the exhibits need to be understood by the public. Indeed, the idea behind an exhibition must be clearly expressed and, if possible, stated to the public. Precisely because it is difficult to reach people who we can’t know in advance and whose age, social and cultural background can only be guessed at, this is both a cultural and, ultimately, ethical obligation. To solve this type of problem, marketing techniques, studies about visitors and all the various tools normally used in market contexts may prove helpful.

However – and this, in my opinion, is the most important point -, the aspect that should always be borne in mind is that exhibitions are experiences. A visit to an exhibition or a museum should always leave a trace in the visitor. Whether it is a cognitive, aesthetic or emotional experience, it should remain in the visitor’s memory. In this sense, it should always be remembered that exhibitions are held in physical spaces and that visitors are made of flesh and blood: in other words, all the necessary conditions – including sensorial, spatial and climatic conditions, supports and facilities – are needed to ensure that the experience is not just instructive and exciting but also comfortable at a purely environmental level.

Besides the exhibits, which, as we have said, represent the “distinctive feature” of exhibitions, each display is accompanied by other elements – intended to accompany visitors and provide them with informative and interpretative tools of varying nature – which are of great importance for the purposes of communication.

We can identify four main groups of elements.

The first regards everything that involves welcoming and looking after visitors and guiding them through the exhibition space, from the ticket office, to the plan of the exhibition areas to the signs for routes and facilities; all these features help visitors to move around the real space of the exhibition or the museum.

The *second* regards everything related to the information about the exhibits. Typically, the main element consists of captions and panels which, for example in an art exhibition, contain the basic information for learning about what we are observing: the name of the artist, his or her date of birth (and, eventually, death), the title of the work, the year it was made, the techniques used, the collection from which it comes if it is a loan or a donation. In most cases, especially in institutions which are most aware of the need to "internationalise" their exhibits, captions and panels will be in the language of the country where the exhibition is held as well as in English.

The *third* gathers together all those elements which help visitors to contextualise and decode the meaning of what they are seeing: this category includes the catalogue, information panels, multimedia products, web pages etc.

Lastly, there is a *fourth* category: there is a whole set of tools of both an informative nature but also of public interest which have the aim of publicising the exhibition or museum and promoting visits in the local area (and further afield). This category includes the typical tools of any advertising campaign: posters, leaflets, brochures, flyers, advertising banners and so on. Other examples falling into this category are even more purely commercial objects linked to *merchandising* such as gadgets, postcards, t-shirts and souvenirs in museum shops; they represent both a source of advertising and an opportunity to raise money for the organiser.

The above points should help to explain why an exhibition, in itself, is a multimedia "event" in the sense that it uses a range of different supports (objects, fixed and moving images, words, etc.) and tools. It goes without saying that the advent of multimedia in the strict sense of the term, i.e. hypertexts and the web, has contributed enormously to the possibility of communicating contents, as well as providing a potentially infinite contribution in terms of both quantity and quality. Nevertheless, it should be emphasised that these multimedia tools can only enrich the experience of the visit and are not a substitute, since the visit is chiefly linked to the relationship between the visitor and an object within a physical environment. For the purposes of this paper, we will make a general distinction between the use of offline and online technologies.

The use of offline technologies provided an important addition to museums and represents the first step in the introduction of new technologies to museum communication. To summarise, these technologies have made an essential contribution, especially in terms of "quantity", for carrying out the following functions:

- Creating a large data archive, such as virtual catalogues of museums, where it is possible to store images, texts, sounds etc., making all the information immediately available to visitors within the confined space of a PC.
- Simulating processes, such as in the simulations used primarily in science museums to reproduce phenomena and experiments which cannot be carried out in the physical and spatial context of the museum.
- Representing/reconstructing worlds, primarily through virtual reality, whether immersive or not, enabling visitors to gain a spatial experience of monuments

and structures that no longer exist (one just think of the importance of these applications in the field of archaeology).
- Establishing new relationships/connections, i.e. linking – at a thematic level – various cultural objects, making it possible to discover new ties between spheres and fields of knowledge which have often been considered to be distant from each other or even antithetical (e.g. art and science).

The situation changed radically during the nineties with the introduction of the Internet. The web provides a dialogue box between the museum and the outside world and the use of the web helps to establish a new relationship between the museum and its visitors. Due to its distinctive features, the computer revolution has led to innovations in terms of "quality", as can be seen through the following examples:

- Production has become confused with consumption. The web has led to the emergence of the *prosumer*, who is part producer and part consumer of the contents. Web surfers can actively take part in defining contents, display choices, debates and discussions about the contents of exhibitions. Alternatively, they can build their own personal "museum", as in the case of web art.
- New forms of intelligence emerge: the collective intelligence theorised by Levy and the connective intelligence theorised by de Kerckhove postulate a new dimension in which knowledge begins to circulate and becomes collaborative. In this sense, museums become nodes in the web since the contents they make available to the public represent "resources" of a meta-encyclopaedia and, simultaneously, virtual places in which to express oneself.
- The senses are enhanced and acquire spatial and temporal extensions. This is true both in terms of contents (the distance use of online museum contents) and the management of more practical aspects, but are just as important, as in the case of marketing and communication, in the life of a museum.

To conclude this brief digression, we can say that about 40 years after the initial experimentation related to the use of computers in museums, the most distinguished scholars of the phenomenon seem to have reached an agreement on the following points. Firstly, the "distinguishing feature" of museums is the experience that takes place within a physical space, a real architectural space which acquires increasing importance and significance (in a symbolic sense as well). Secondly, the dream or nightmare – depending on the point of view – of the "virtual museum", which was supposed to replace actual visits, is in crisis: the number of visitors to museums and temporary exhibitions is constantly growing and the increase appears to be relentless. Lastly, the current trend is for new "real" institutions which can communicate with the "digital" dimension and where virtual reality involves a constellation of services which accompany, prolong and complete the experience of the concrete, physical visit to the museum.

However, a museum and/or an exhibition do not just exist as a result of the objects they contain or the technological supports – both new and old – that contribute to the visit. A role of fundamental importance in the process of museum communication is assured by people. We shall not devote space here to the people who look after the museum contents, the so-called "curators", but rather to those that have a direct contact with the public. The main roles involved in this process regard those responsible for the reception of visitors – attendants, ticket office assistants, etc. – and especially the guides who present exhibits to the public.

The role of the museum guide has changed enormously over time. While guides used to have the task of conveying notions by simplifying the contents on display in museums and exhibitions, the current tendency – especially in science museums – is for guides to be "active"; their principal task is to stimulate visitors' curiosity, encouraging them to find answers rather than to ask questions.

In this sense, it is no coincidence that other terms for the role of guide have often been coined: facilitator, entertainer, helper, etc.

It should be stressed that the central role and importance of human mediation in museums is universally recognised nowadays, while studies and research demonstrate that when communication is provided only through traditional methods, it leaves less of a trace on the visitor's memory and learning.

Brief mention should be made of the use of theatre, which is becoming an increasingly popular means of conveying information about museum contents. Its use in museums of all types, ranging from art museums to archaeological, technical and scientific museums etc., has a series of advantages. Firstly, it offers the possibility of giving a dynamic interpretation (in time and space) of the contents. Secondly, due to its flexibility, it provides the opportunity to establish a dialogue with the visitors/spectators, both in terms of the questions and specific aspects of the themes involved (e.g. additional information about works, authors or artists, etc.) and also in terms of controversial issues, especially with regard to science museums. Last but not least, the theatre creates an emotional context which is of enormous importance in cognitive processes, both for children and adults.

The concept of "event" is sufficiently broad to encompass a wide range of possible encounters between the visitor and the museum. We shall merely observe here that together with the "loss of sacredness" of the museum as a place and institution, the people and organisations that run museums in various capacities have opened their doors to host a range of events to enliven their premises: it is no longer unusual for shows, society events, fashion shows, receptions and parties for children to be held in museum premises. On the one hand, this contributes to museums' need for self-financing, which is increasingly necessary in a phase of cuts in public spending on culture; on the other hand, it helps to raise awareness of the museum among the general public who might otherwise not be attracted to the standard events held at museums – i.e. exhibitions.

To conclude this chapter, it is worth recalling that behind each of these activities there are real people who have well-defined professional skills. This theme is too long to be explored in depth here; suffice it to say that the professional profiles linked to the world of museums are constantly diversifying and increasing, partly due to museums' need for self-financing (professions related to marketing and fund raising), the rise of new information and communication technologies and the increasingly "social" nature of museum activities. The terms that used to be associated with museums (curator, conservator, etc.) are gradually disappearing and are being replaced by new roles and skills.

Part II
Science Centres

The Evolution and Worldwide Expansion of Science Centres

Vincenzo Lipardi

Introduction

If we were to photograph the world of science centres today, we'd have a picture of a complex reality with various forms, a movement being continually expanded on and organized at the International level with regional networks that put thousands of small, medium and large centres together. It is a community made up of tens of thousands of men and women, passionate about their jobs, experts in science communication, skilled in the management of "museums" visited each year by thousands of millions of visitors. It is an enormous cultural network, a network of good scientific information and, at the same time, an important economic segment in the cultural tourism market.

Science centres have also developed extraordinary capacities in the field of informal education: the activities produced inside their educational laboratories, due also to the Internet, make up an important part of the education offered by each good professor of sciences. Science centres have become strategic hubs, combining schools, research centres, universities and scientists. Finally, science centres are becoming more and more a tactical instrument used to strengthen the flourishing relationship and dialogue between science, society and the public opinion, continually contributing to the creation of a true scientific citizenship, indispensable for understanding the modern world. Science and especially modern technosciences have become too much important to be left to the market. The role that citizens play in transmitting their needs, interests and ideas is fundamental.

The physicist Vittorio Silvestrini, father of Città della Scienza in Naples, uses the following words when presenting the centre's mission. In the book *La risorsa infinita* (*The Never-ending Resource*), Silvestrini defines "culture" as a heritage of

V. Lipardi (✉)
Chief Executive Officer of Fondazione Idis-Città della Scienza, Naples, Italy
e-mail: lipardi@cittadellascienza.it

A.-M. Bruyas and M. Riccio (eds.), *Science Centres and Science Events*,
DOI 10.1007/978-88-470-2556-1_5, © Springer-Verlag Italia 2013

knowledge, made of common and shared values possessed by a specific human community.

> Once primary needs are satisfied (hunger, thirst, health, etc.), culture (together with the quality of the physical and social environment) is also the parameter on which the quality of life mostly depends: in this sense, culture is itself a relevant primary good.

Yet in the twentieth century, humanity has experienced a real economic, cultural and moral revolution, entering into the era of the "society of knowledge". Silvestrini affirms:

> It is above all a cultural revolution, tracing its origins in the modern scientific revolution started by Galileo and leading to the great scientific theories of the 19th and 20th centuries: thermodynamics, electromagnetism, relativity, quantum mechanics, evolutionism, and genetic biology. So, a primarily scientific cultural revolution takes place, affecting every other cultural field. Since the first decades of the 20th century, it is no longer possible to study philosophy (metaphysics, aesthetics, epistemology, etc.), or any other cultural form, disregarding scientific achievements.

This is the cultural context in which science centres work, renewing the cultural climate in which they were created and strengthening the need for a scientific citizenship that has slowly begun, in the last decades, to establish itself in a "karstic way" into society.

What really are science centres, how did they come about, what is their role in the "society of knowledge" and, finally, is there a common goal they want to achieve?

From the "Toronto Declaration" to the "6th Science Centre World Congress" in Cape Town

The 5th Science Centre World Congress tried to answer this last question in 2008, approving the "Toronto Declaration". After a long series of meetings at the international level: the First Science Centre World Congress took place in Finland in 1996 thanks to the enthusiastic and competent work of the director of Heureka, Per-Edvin Persson, one of the most influential people in the Science Centre "movement"; the congress was then held every 3 years in India (1999), Australia (2002), Brazil (2005) and Canada (2008).

Signed in 2008 by ECSITE for Europe, ASTC for North America, SAASTEC for South Africa, various Asian networks (India, China, Pacific) and by Red Pop for Latin America, the "Toronto Declaration", promoted by Lesley Lewis, the brilliant CEO of the Ontario Science Centre, represents the desire to create a common framework and platform between scientific museology in order to play a role in the globalised world. As the declaration emphasizes:

> Each year 290 million citizens actively participate in the exhibitions, programs, events and outreach initiatives organised by 2,400 science centres world wide. Science centres stimulate curiosity and develop enquiring minds. They change people's lives, influencing their attitudes and thinking. Research shows that science centres demystify science, conveying

> its beauty, showing its necessity and making it accessible to the general public. They foster positive attitudes towards science, help people to appreciate the context of scientific advances and to understand how science affects their lives.

Among its main themes, the Declaration also states:

> In 2008, science literacy is as important as other forms of literacy and numeracy. It is also a powerful tool for social inclusion. Science centres have relevance to all sectors of the population and have become important meeting places for science and society. They operate across geographical, economic, political, religious and cultural boundaries. They impact the well-being, education, achievement and skills of current and future generations. They are safe places for difficult conversations. Around the world today, science centres are highly-visible and trusted hubs of activity, dialogue and discourse about science and technology; they support the skills needed for effective problem-solving, creativity, innovation, critical thinking and decision-making, therefore enhancing lifelong learning of science and technology. They are important resources for the formal education system and contribute to strengthening the knowledge bases of their respective societies; they influence the motivation of students, the learning process and their career choices; they empower teachers, introducing them to more effective ways of teaching science, mathematics and technology. They create important platforms for the increasing number of virtual visitors engaging with science – and with each other –using on-line and digital technologies. They influence research and museology related to science communication, education and engagement. They are trusted places of inclusion and equity where the public can actively engage with critical issues which affect society.

The Declaration arose from a trend that has become unstoppable in terms of the "technical" and "museal" purpose of a museum, to a "social" and "political" function that is naturally part of all cultural institutions, but that are emphasized in the face of challenges brought about by modern times and globalization:

> They form strategic partnerships to help address important local, national and global challenges. At the Fifth Science Centre World Congress, delegates shared expertise and developed common agendas for action. A new era of global cooperation respecting local cultures is needed if we are to successfully live on a planet with diminishing resources and significant environmental challenges. Science centres can be a powerful force for good. Children who attend our science centres are growing up in a rapidly changing world and can become critical "agents of change" so that everyone can have a better future. Teens and university students who participate in science centre programs are tomorrow's leaders and decision makers. Adults who visit our centres and get involved re-engage with science and become better positioned to understand the context of scientific discoveries and contribute to dialogue on topics such as climate change, human health, the need for renewable energies, water shortages and HIV/AIDS.

Finally, the Declaration makes a solemn commitment:

> We commit to work together to overcome cultural, physical, social, economic and geographic barriers to engage and connect people through science. We will actively seek out issues related to science and society where the voices of citizens should be heard and ensure that dialogue occurs. We will work together to identify how science centres can contribute to the achievement of the United Nations Millennium Goals. We will search for funding and mechanisms to create a better future for all through global engagement with issues of local, national and global relevance, including environmental awareness, science education and innovation. At the Sixth Science Centre World Congress, to be held in Cape Town, South Africa in 2011, we shall assess the extent to which we have, as individual institutions and collectively, moved forward on the goals established here.

How did we arrive at the "Toronto Declaration"? I personally think we can find the answer to this question in the creation of science centres, knowing that, like in every other social institution, museums evolve constantly and it is more than likely that in 20 or 30 years we will need to have another look at all the interpretations we have made, comparing them to the cultural actors that will be radically changed with respect to today.

The Toronto Declaration was recalled in 2011 by the 416 delegates attending the 6th Science Centre World Congress in Cape Town who continued the dialogue with the previous world gatherings and formulated a new declaration to ensure that they continue to play a constructive role in addressing global issues at the interface between science and society. The Cape Town Declaration assessed the progress done in the past three years and resolve leaders to encourage the establishment of science centres and museums in parts of the world where they are lacking, to support a policy investment in science, technology and innovation in response to the global economic and financial challenges, to work together to ensure that they share their joint experience and knowledge of the most effective methods.

What Are Science Centres?

Luigi Amodio takes a paragraph to explain the passage from traditional museums to Science Centres in this book. In addition to his considerations, which you have read before, I would like to add Hilde Hein's view about this new generation of museums that breaks away from traditional culture.

> Museums usually house static displays to be admired from a respectful distance. Even natural history collections, the primary content of science museums, tend to require only a passive and reverential appreciation of the odd and various specimens someone has painstakingly assembled. Such exhibits can instil wonder at the rich diversity of the universe, but also tend to encourage for the brilliance of those few scholars who have been able to unravel its complexity. Science museums often glorify scientists more than teaching museum goers the practice of science. Visitors are invited to admire the accomplishments of others, but they are not motivated to think that they might go and do likewise.

Few years ago Pietro Greco wrote:

> The science centre is the answer that a part of the scientific community has given to the positive science crisis. After Hiroshima and after the acquired knowledge of the limits of development founded on a proprietary conception of nature, the idea of science – of a science that has now received social prestige and concrete funding like never before – has separated (or appears to have) from the idea of progress.

The science centres "community" owes much to Frank Oppenheimer, the American physicist who founded the Exploratorium in San Francisco in the 1960s, the same time the Ontario Science Centre in Toronto was opened, initiating a new era of scientific museums. In fact, the need to create a new generation of museums geared more towards the scientific method and "laboratory" work had been argued for some time. Oppenheimer gave this need a firm theoretical basis and a work method that was acknowledged at the international level in a very short time.

Oppenheimer's "museum" is based on his philosophy of science. He says:

> The whole point of the Exploratorium is to make possible for people to believe they can understand the world around them. I think a lot of people have given up trying to comprehend things, and when they've turned their backs on the physical world, they've turned their backs on the social and political world as well. If we give up trying to understand things, we'll all be sunk.

Oppenheimer proposed the creation of a place that offered an approach to science not just able of attracting visitors to marvel and stand in awe before the scientific achievements attained, but that invites them to "make science" themselves through games, experimentation and reproduction of experiences. Science centres try to reaffirm that science is not just useful, but represents a form of culture needed to understand the world. Furthermore, science is for everyone and its nature and phenomena, beginning with the infinitely small in order to arrive at the infinitely large, can be inquired into and understood.

Teaching visitors is not the goal (there are other institutions for that). The goal is to make visitors curious, discovering that everyone, despite social conditions and cultural backgrounds, can come into contact with science, and this is the first indispensable brick to creating "scientific literacy".

In a wonderful book entitled *Science Matters*, physicist James Trefil and geophysicist Robert M. Hazen write:

> Scientific literacy is the knowledge we need to understand public issues. It is an eclectic mixture of facts, vocabulary, general principles, some history and some philosophy. This knowledge is not the specialized stuff of experts. It is the more general and less precise knowledge used in political speeches. If someone can take a newspaper article about genetic engineering or the ozone hole –if they can treat science in the same way they treat any of the other pieces of information- then they are scientifically literate. This definition of scientific literacy may seem quite limiting and even completely inadequate to some scholars. But we are convinced that people who would like for everyone to possess profound scientific knowledge are confusing two important, but very different, aspects of scientific knowledge. The point is that making science is completely different from using science. Scientific literacy regards only the latter.

Science centres are still quite new to us if we think that the first two institutions to use this approach were both opened in 1969, one a few months after the other.

On August 20, 1969, in the Palace of fine Arts, the Exploratorium opened its doors to the public:

> Tourists and curiosity seekers [. . .] discovered the museum largely by accident and went in to see what was going on. There were almost no exhibits inside, only a small band of people industriously at work and a sign that still reads: *Here is Being Created the Exploratorium, a Community Museum Dedicated to Awareness.*

At the same time in Toronto, Canada, the planning for a new science centre began in 1961 during Toronto's massive expansion in the late 1950s and 1960s. In 1964 the famous Toronto architect Raymond Moriyama was hired to design the site. Building started in 1966 with plans to make it a part of the city's 1967

Canadian Centennial celebrations. It was officially named the Centennial Centre of Science and Technology. However, construction was not complete in 1967, and the Ontario Science Centre (OSC) did not open to the public until two years later, in September 1969. The OSC was famous around the world for its hands-on approach to science, which was later imitated by Detroit's Museum of Science and Technology. Unlike the traditional museum where the exhibits are there to be looked at, the majority of the exhibits at the OSC were interactive and many others showed live demonstrations (i.e. metalworking). The Communication Room was particularly well-loved, containing a number of computerized displays, including a very popular tic-tac-toe game run on a PDP-11.

Pietro Greco recalls that physicist Jean Perrin had embarked on this same path when he founded the Palais de la Découverte in Paris in 1937:

> Perrin's museum tried to involve the general public, like Bacon's idea and with other structures in the 1600s, proposing an approach to science that differed from "positive museums". His goal was not to offer a utilitarian and sometimes triumphant image of science, but to propose cultural value for what it is and creativity as an approach to science. Perrin was probably ahead of his times, but the Exploratorium that the physicist Frank Oppenheimer founded in San Francisco in 1969 is the evolution of his museum.

Philip Morrison, in the foreword to Hilde Hein's book *The Exploratorium: The Museum as a Laboratory* writes in reference to the Exploratorium in San Francisco:

> The hope behind this display of friendly marvels is the unity of human beings. A unity that lies below all their differences, that rests in their common ability to share in reason and delight. The egalitarian hope is not new in science. Perhaps it rises in all the times of transition, as it appeared in the days of the tireless traveller Alexander van Humboldt, who took his thermometer and his search for unity to the cold current of Peru. He was a democrat in the service of the Kings, a man between eras; he journeyed romantically afar but he always took along his instruments, the critical spirit of the Enlightenment. Oppenheimer's vision was that human understanding will cease to be an instrument of power, benefiting the few, and will instead become a source of empowerment and pleasure to all.

Hilde Hein says that:

> Oppenheimer sought to bring about a museum in which people would directly experience and manipulate things, instead of being told about them. The public was to interact with objects as an experimental scientist does in the natural world or in a laboratory. The museum was to teach that the subject matter of science is all around us and that it can be understood by all. It was to remove science from the exclusive domain of experts, to demystify it, and to restore it to the common sphere.

Since then science centres have sprouted up all over the world, many traditional museums have drawn on these same ideas, introducing new kinds of activities to their collections based on interactiveness and communication with the visitors. We would now like to take a quick trip around the globe, through the regional networks that organize science centres in order to trace the development and impact that science centres and museums have had.

Worldwide Networks

The Toronto and Cape Town Declarations were signed by seven associations that represent an important part of the macro-regions on our planet. Meanwhile, other regional networks have been created. The creation and development of this movement is due to the hard work of many people, yet the Association of Science-Technology Centres (ASTC), the North American network, has played an extraordinary role. The association, having initiated a movement whose development quickly spread to the four corners of the globe, is well-rooted in North America and tied to US political and cultural institutions and has always operated with a global vision. Thus today ASTC, despite the presence of other strong regional networks that we will cover in the next few pages, finds it limiting to call itself a regional network and, instead, refers to itself as the Association of Science Centres. This has obviously not limited its availability to collaborate with the others. The other regional networks are reported below as they are described on their websites and in official documents.

The Association of Science-Technology Centres (*ASTC*) founded in 1973, is the network of North America gathering now nearly 600 members from 40 countries. Members include not only science centres and museums, but also nature centres, aquariums, planetariums, zoos, botanical gardens, and natural history and children's museums, as well as companies, consultants, and other organizations that share an interest in informal science education. ASTC and its members were the founders of this movement in the XX century. As announced in its website, it is:

> An organization of science centres and museums dedicated to furthering public engagement with science among increasingly diverse audiences. ASTC encourages excellence and innovation in informal science learning by serving and linking its members worldwide and advancing their common goals. Through strategic alliances and global partnerships, ASTC also supports science centres and museums in proactively addressing critical societal issues, locally and globally, where understanding of and engagement with science are essential.

The Asia Pacific Network of Science & Technology Centres (*ASPAC*) was formed in 1997 to facilitate communication and cooperation among centres, museums and other organisations which use interactive approaches to encourage excellence and innovation in informal learning and the public understanding of science and technology in the Asia Pacific region. ASPAC today has 59 members and draws its membership from 20 countries and administrative regions in Asia. Members include here again many forms of organizations: science centres, science museums, children's museums and exhibit design and fabrication firms.

The Network for the Popularization of Science and Technology in Latin America and the Caribbean (*Red-POP*) was established in 1990, in Rio de Janeiro, inspired by UNESCO's Science, Technology and Society Program. Currently the Red-POP has over 80 members from more than 15 countries in Latin American and the Caribbean and it has relationships with science and technology popularization centres in many countries throughout the world. There are 102 science centres in the region which, according to the size, are divided in 13 large science centres, 20 medium and 69 small or very small. 73 of the 102 science centres are located in South America, although 58 (80%) are small or very small. Central America, Mexico and the Caribbean have 29 science centres. Most of them (18) are large and medium science centres. The statutes clearly announce its mission of providing "a better understanding of the role of science and technology in development – understood as a comprehensive, endogenous and man focused process – is crucial". Difficulties in integrating scientific and technological principles, information and capabilities to Latin America's daily economic, social and cultural activities reveal the need to stimulate the study, debate and interchange of science and technology popularization programs and activities. As for scientific and technological knowledge, its diffusion in developing countries is imperative. On the other hand, the different science and technology popularization programs frequently face limits to their scope due to the countries' generalized budget restriction. To date, there is no regional mechanism to boost national efforts. Red-POP seeks to mobilize national and regional potentials through different cooperation mechanisms to intensify the popularization of science and technology in the region.

The National Council of Science Museums (*NCSM*), an autonomous society under the Ministry of Culture in India, was formed in 1978. Today it administers 27 science centres/museums and planetariums spread all over India located in different sites. It provides catalytic support to other organizations for setting up new science centres within and outside the country. Continuous upgrade of display materials and technology happens in the Council as a part of its R&D activity. It has supported and facilitated the development and expansion of science centres in many regions of the country.

The Southern African Association of Science and Technology Centres (*SAASTEC*) has a long term mission in creating a technologically-based Southern African society which will ensure a competitive edge in the global economy. Most of the currently active science centres who are members of SAASTEC are located in South Africa. Botswana now has a science centre at BOTEC, and Namibia is actively pursuing plans for science centres. Membership provides access to the resources of SAASTEC and its members, in terms of networking, sharing ideas, concepts, knowledge, experience, training and more. Despite only be founded recently, SAASTEC is recognised by the South African Department of Science and Technology as a representative of interactive science centres in the country. SAASTEC held the Science Centre World Conference in 2011 demonstrating the interest of Africa in the development of science centres as centres of excellence and innovation in non-formal science education.

We must not forget the *Chinese Association of Natural Science Museums* that is part of the China Association for Science and Technology, created in 1979 by the Beijing Museum of Natural Sciences that is playing an increasingly more important role in the field of museums today.

Our brief trip around the globe arrives in the Mediterranean basin. *The North Africa and Middle East Science centres network (NAMES)*, was launched in January 2006 during a meeting organized by the Bibliotheca Alexandrina

Planetarium Science Center. The purposes and goals of NAMES are to be achieved through excellence and innovation in informal education and by offering hands-on activities where the audience can indulge in participatory learning. The goal of NAMES is the popularization of science throughout the region by enhancing public understanding and involvement in science and scientific culture among an increasingly diverse audience. The purpose of creating NAMES is to foster cooperation between existing science centres and museums in North Africa and the Middle East with the aim of benefiting from all available resources and experiences within the region to further enhance the role of all member establishments. The network also aims to help the establishment of new science centres and museums in all countries of the region. There is a contradictory relationship between the Arabic Science Center and Israeli's Science Center, whose ties are "frozen", a fact that deeply concerns all the regional networks, and in particular ECSITE.

Last but not the least, ECSITE, the network of European museums and science centres, that I had the honour to chair from 2007 to 2009. The *European Network for Science Centres and Museums* (*ECSITE*) was founded in 1989. As one of ECSITE's well-known Presidents and one of the most recognised European leaders in the science centres movement, Asger Hoeg writes:

> The idea of creating a European network of science centres and museums was brought about by Joel de Rosnay, director of Development and International Relations of La Cité des Sciences.

Rosnay invited European delegates to a meeting during the ASTC conference in Boston on October 17, 1988. Ten people participated in the meeting. The Boston group decided to form ECSITE and sent invitations to a founding meeting at la Cité des Sciences et de l'Industrie on January 9, 1989. In that meeting, the participants nominated Richard Gregory from the Bristol Science Centre as the first President of ECSITE and Melanie Quin the General Secretary. ECSITE was thus conceived.

Today ECSITE links science communication professionals in more than 400 institutions in 50 countries through projects and activities, facilitating the exchange of ideas and best practices on current issues. Members include science centres and museums, science festivals, natural history museums, zoos, aquariums, universities, research organizations and companies communicating and engaging the public in science through accessible, interactive exhibits and programs. Science centres and museums are trusted by the public as independent sources of information, since they provide opportunities to listen to different points of view about scientific issues. ECSITE member institutions attract over 30 million visitors in their venues and many millions more through their websites. Science centres and museums provide students with educational activities that complement school curricula and offer training and professional advancement programmes for teachers. Science centres and museums provide a common ground where scientists and the public can meet and discuss controversial and contemporary issues about science and technology, a crucial element for the economic and social development of Europe.

Building a Bridge in the Mediterranean Area

More than in any other region of the world, the Euro-Mediterranean is evolving rapidly. A new age of hope for empowerment and freedom across the Arab world has been inaugurated, which is inspiring social mobilization everywhere including Europe. Our movement can play an important role in the establishment of a new and innovative Euro-Mediterranean vision. This means not only to act in our limited local areas but also globally, because science and technology are the result of supranational exchanges and cross-fertilisation that make the old idea of national boundaries absolutely obsolete. When the scientific communities of different cultures and countries work together, it is possible to attain shared solutions. Science is a universal language that is capable of uniting people and is the instrument that creates answers to the need for economic and cultural growth through the security and peace of the people in the region.

Therefore the two networks NAMES, ECSITE and some of their members (Città della Scienza Naples, Biblioteca Alexandrina Planetarium Science Centre, Scientific Centre Kuwait, Museo delle Scienze Trento, Parque de las Ciencias Granada, Natural History Museum London, Universcience and Universcience Partners Paris), have recently launched a project to create a Summer School of Scientific Communication for the Euro-Mediterranean and the Middle East area. A memorandum of Understanding was signed in June 2012 during the ECSITE Conference in Toulouse by the two Presidents: Robert Firmhofer for ECSITE and Mijbil Almutawa for NAMES. The Summer School will be addressed to middle profile staff of existing science centres and museums and to new comers from other organizations (universities, local entities, associations) who want to establish new activities in the field of science communication or to develop new science centres.

Main objective of this project is to reinforce the capacities of development of science centres, but also to strengthen dialogue and exchange among these institutions with the scope to build a growing community of actors in the region and to maximize their potential impact as agent of change for societies.

The New Frontier

Opening the ECSITE Annual Conference in 2007, professor José Mariano Gago, the Portuguese Minister of Research, proposed a challenge. He asked if ECSITE was ready to act on the political debate about revisiting and revitalizing the EU's Lisbon Agenda and if they were conscious of the strength and importance of the science centre movement. ECSITE established three new directions to move in during this conference:

> The first, to facilitate cooperation among its members, establishing guidelines, sharing expertise, disseminating best practices, encouraging collaboration and developing training programmes.

> The second, to capitalize on the European dimension of the network with the goal of contributing to the creation of a European scientific citizenship. This not only means that Europe is above all a geographic and political place that unites citizens, having them work and think in a supranational dimension, but also, and above all, it is a common area with a common cultural feeling (especially among the new generations) and gives way to some contradictions in modern times, from digital diversity to the role of science and technology in the daily lives of its citizens. In addition, it was decided that one of the aims in the first two years after the conference would have been to strengthen the relationship with European institutions and, at the same time, to assure that its members develop the skills necessary to submit European projects, create an event within the European Parliament with European institutions and focus on the topic of science and society, bringing together experts, political representatives from European countries and the officers and decision makers of European institutions.
>
> The third objective was to reinforce relations and partnerships in the international arena. ECSITE represents European science centres and museums. Maintaining active contacts with network organisations in other parts of the world, ECSITE should also promote important international actions on global issues such as climate change, energy saving and renewable sources.

There were numerous responses to these indications. From the political point of view, the most important initiative was most definitely the conference organized on February 18, 2009 within the European Parliament, which gathered representatives of key institutions both from inside and outside the field of science centre communication in order to define their respective roles in the Lisbon agenda.

Policy makers, science communication experts, stakeholders and practitioners met inside the European Parliament to address the question, "what can we learn from the work accomplished to achieve this goal and what does the future hold in store for the field of science communication in Europe?".

More than 200 people participated at the event, opened by Catherine Franche, the Executive Director of ECSITE, and by myself in the quality of President of ECSITE. The scientific and political success was guaranteed by the reputable speakers present, among them Janez Potocnik, European Commissioner for Science and Research, Philippe Busquin, MEP and Chairman of STOA, Luc Van den Brande, President of the Committee of the Regions, Josè Manuel Silva Rodriguez, Director General of DG Research at the European Commission and José Mariano Gago, Portuguese Minister of Science, Technology and Higher Education.

The same year in Copenhagen, on December 15, 2009, ECSITE held an official side event within the COP15 conference in conjunction with the World Ocean Network, ATSC, the WWF Artic Centre. This event took place within the framework of the ACCENT project, an EU cooperation project that aims at strengthening the efforts of European science centres by engaging citizens on climate change issues. This event was incredibly important for having brought the voices of operators and NGOs active in the field of scientific communication before the uneasiness of politics that would have brought about the significant failure of the summit.

ECSITE also coordinates large European cooperative actions under the sixth and seventh Research Framework Programme. In the field of science education, PENCIL analyzed pilot projects in 14 science centres with the research work of King's

College and the University of Naples in order to identify the criteria of innovation and quality to transfer them from informal learning to formal learning practices. PENCIL assessed the role of science centres and museums in providing students with educational activities that complement the school curricula and offer training and professional advancement programmes for teachers.

With projects GAPP and later TWIST, ECSITE aims at raising awareness about the role and representation of women in science and technology throughout science centres and museums in Europe. They were targeted at young people, their teachers and parents as well as the general public, focusing on the outdated stereotypes with regards to societal roles for men and women and existing obstacles in career paths.

ECSITE has begun in 2011 PLACES, a 4-year European project that establishes and develops the idea of a European City of Scientific Culture and has led to other European networks: EUSCEA (the European Science Event Association), ERRIN (the European Regions Research and Innovation Network) and the University Pompeu Fabra in Barcelona. PLACES involves 67 science centres, museums, festivals and events, each partnering with local authorities, and 10 European regional networks and is expected to generate a wealth of new knowledge, contributing to the development of the whole field of science communication and impacting citizens and policymakers with the social, cultural and scientific added value of scientific culture.

Looking Towards The Future

Only 40 years have passed since the first science centres were created and both the community and its ideas have grown, developed and transformed. It is obviously very difficult to predict the future considering how the international economic crisis has so quickly weakened collective trust in unlimited growth and in spreading areas of wellness worldwide, making everyone think twice about their own role and purpose. However, the resource of knowledge, as expressed in the first few pages, is boundless and the future of our species rests in its hands.

Today, as international speculation is attacking the Euro and Europe, the petrol disaster that destroyed flora and fauna reminds us, on one hand, that our planet is tired of suffering the consequences of the high-handedness of human beings and, on the other hand, profit must not be the only goal of mankind. Likewise, the eruption of the Icelandic volcano, which suddenly put the entire global air transportation system in crisis, questioned our finiteness and, all things considered, weakness before nature.

The future is thus complicated and the need for a new model of "sustainable" development, as we say today, is no longer a request from our planet's more humanistic side, but a pressing need that our world leaders must respond to in uncertain times. The society of knowledge has re-designed the concept of time and space and that which is true for goods and financial markets is beginning (or rather, it always has been, but public opinion has only begun to be conscious of it now) to be true for human beings and for political decisions as well.

What is certain is that we must build a more united and interdependent world, more supportive, more scientifically literate and more associated with the network. In essence, science centres have a lot to do and a lot to think about.

The proposal put forth by the IPC (International Program Committee), the committee working on the Science Centre World Congress demonstrates how widespread this idea has become. One of the improvements proposed was to transform the conference into a World Summit which aims at advancing our field at a higher strategic level by generating a long-term action plan and developing a declaration which is action-oriented and involves and consults external agencies (UN, etc.) in its planning and programs. Fundamentally it aims at strengthening the nature of the community's movement starting with, despite the different viewpoints that exist among the many participatory institutions, the many things we all share a passion for science and knowledge, devotion and respect for our visitors (especially the young ones) and the desire to build a comprehensive, sustainable and fair future.

Part III
Practical Guidelines to Approach Science Communication

The Cooperation Project for a New Science Centre in Owerri, Nigeria

Anne-Marie Bruyas

Between 2008 and 2009, Città della Scienza, together with the Pontifical Council for Culture and ASCO, a group of young students from La Sapienza University of Rome and Nigerian Universities, promoted the first Science Festival in Owerri, which took place from April 26 to May 4, 2009.

This was the first practical initiative in that region, the first step of a long process towards the promotion and the diffusion of scientific culture to all Nigerians, and particularly young Nigerians, with the fulfilment of a long-term science centre. The festival also aimed at increasing the awareness of local and international institutions about youngsters interest in science as a crucial condition for the future development of the African continent. Science and technology are indispensable tools to enable people to take their destiny into their own hands, implementing a strategy of empowerment. In this sense, the festival has also been a real test of the importance and value of scientific communication throughout Africa in helping future generations to become more aware and independent. More specifically the project aims to respond to the following goals:

- To spread a scientific culture based on experimental sciences among the local population;
- To attract the youth towards scientific careers, stimulating their curiosity and enhancing their skills;
- To improve the quality of scientific education, promoting informal learning methods and the practice of scientific laboratories in the school and with the university system;
- To build capacities of local staff as future science communicators and educators (students, scientists, local officers etc.).

A.-M. Bruyas (✉)
International Relations and Projects, Fondazione Idis-Città della Scienza, Naples, Italy
e-mail: bruyas@cittadellascienza.it

A.-M. Bruyas and M. Riccio (eds.), *Science Centres and Science Events*,
DOI 10.1007/978-88-470-2556-1_6, © Springer-Verlag Italia 2013

Fig. 1 Visit at the Imo State University

In order to develop sustainable activities we proceeded with the active involvement of local actors by three progressive steps: the context analysis, the training of local staff, the setting up of the science festival.

Context Analysis and Mission in Owerri

This first phase consisted of two interconnected actions: the analysis of the local context in its different aspects (population, economy, education and training, scientific development) and the organization of a first mission in situ, from the 23rd to the 29th of July 2008. It was essential for the promoters of the projects (Pontifical Council of Culture, Città della Scienza and ASCO) to identify and involve in the implementation of the action the likely local partners among the different stakeholders: scientific community, school representatives, institutions responsible for education, science and culture, international organizations, etc.

The mission in Owerri was an important phase to verify on the spot the availability of the institutional, scientific, cultural and religious forces, either local or national, in order to undertake the real achievement of the project. During the mission the delegation had several encounters with the representatives of various universities: Federal University of Science and Technology, Imo State University (Fig. 1). Federal Polytechnic Nekede Owerri, College of Agriculture, College of Education, Abia State University, Nnamdi Azikiwe University, School of Nursing Midwifery; the Imo State Government and specifically the Councillorship of Education and the Councillorship of the Public Function; the Italian Embassy in Abuja; the delegates of International organizations (UNESCO and the European Union); the Archdiocese in Owerri.

The situation in Owerri, although very difficult, both for the condition of the country and the proximity with the Niger Delta, suited the proposed operation, also considering the possibility to create a synergy with the local public institutions, particularly the Government of the State and several universities. These institutions declared their availability to support the initiative.

The Catholic Church is very popular in the Imo State and has credibility and consent in wide layers of the population as well as a good relationship with the academic community. During the mission and the meetings with exponents of the Nigerian Church, it has been possible to verify the high cultural and human quality of the local leadership.

Also the contacts in Abuja encouraged going ahead with the operation. Particularly UNESCO was planning a territorial system of science centres to be developed in the six geopolitical areas of Nigeria. The European Union delegation in Nigeria was also interested in the initiative because all their development programs in Africa are based on raising awareness in the local population. Finally, the Italian Embassy showed its interest to follow the initiative in the following months and sped up the delegation to contact the Ministry of Foreign Affairs in order to understand how to insert it in the cooperation participation in Africa. All the outcomes convinced us that our plan for the achievement of a science centre was a peculiar and innovative initiative in the sub-Saharan regions.

Training Program for Nigerian University Students

The second important step was the organization of a training course to be held from 3 to 17 September 2008. Nine Nigerian university students had to attend the course because they would have been in charge of the implementation of the Owerri Science Festival in the following months. According to the organization of the project, beneficiaries of the training could also be: post graduate students, scientific researchers or professors, science teachers, public officers (Fig. 2).

The course included four main topics: the planning, the communication and the promotion of a scientific event, the techniques of communication with the public and the didactics. This last topic played an important role in the course and was divided into five different areas for the Science Festival: a main area with a "hands-on" exhibition about the main principles of classic physics; and four thematic areas for science labs – environment, energy, health and food, music – as previously chosen in agreement with the Local Scientific Committee.

The full programme of the training course and the lessons are reported in Table 1. It also included a visit to the laboratory of the association Scienza Viva in Calitri with the collaboration of the Maffucci School Institute. An entire day was dedicated to the EASW (European Awareness Scenario Workshop) in which the students discussed with the experts the above main features of the Science Festival and the future Science Centre (Fig. 3).

Finally an Italian language and culture lab was also organized in collaboration with the European Centre Information Citizenship (CEICC) of Naples. The visit to the historical centre of Naples and to the National Archaeological Museum and a trip to Capri were the occasion for a profitable intercultural exchange.

Fig. 2 Nigerian participants to the training course

Fig. 3 Practical workshop with hands-on exhibits

Setting up of the Owerri Science Festival

The third step was the organisation of the Science Festival in Owerri at the end of April 2009. The implementation of a temporary event like the Science Festival represented a real action of scientific communication towards the local public. Its success was also a test on a short period of the interest of the public, the impact on the local system and the efficiency of the organizing local partnership. It has the advantage to concentrate big efforts of communication and advertising in a short period and on the other hand to have limited infrastructure and maintenance costs because it is temporary.

Table 1 Training program for Nigerian students, Naples, Città della Scienza, 3–17 September 2008

3 September 08 **Welcome words** Vittorio Silvestrini, *President of Città della Scienza* Vincenzo Lipardi, *President of Ecsite, Member of the Executive Committee of Città della Scienza* Tobechi Anyadike, *President of ASCO* **Introduction of Nigerian students and presentation of the Owerri local context. Presentation of the training program** Anne-Marie Bruyas, *Responsible for International Relations, Città della Scienza*	**8 September 08** **Promotion and communication tools** Carlo Guardascione, *Responsible of Communication, Città della Scienza* **Science centres visitors needs** Francesco De Matteis, Explainer, *Fondazione IDIS-Città della Scienza* **Communication techniques with visitors.** Davide Petrone and Guglielmo Maglio, *Explainers team, Città della Scienza* **Planetarium and astronomy basics** Alessandra Zanazzi, *Responsible for astronomy activities, Città della Scienza* **Italian language course**
4 September 08 **Concept and history of science centres** Luigi Amodio, *Director of Città della Scienza* **Connection between formal and informal learning** Mario Campanino, *Responsible for Education, Città della Scienza* **Science and society: the role of science and science communication for/in society** Pietro Greco, *scientific journalist* **Italian language course**	**9 September 08** **Visit to the hands-on exhibition and workshop "Square wheels" in Calitri** Pietro Cerreta, *Association Scienza Viva, Industrial High School Maffucci, Calitri (AV)*
5 September 08 **Typologies of activities in science centres** Guglielmo Maglio, *Responsible for exhibitions and visitors activities, Città della Scienza* **Science centres exhibition area** Luigi Cerri and Anna Porro, *Education unit, Città della Scienza,* Fiorella Iuliano, *Explainers* **Exhibition setting up and backstage.** Maria Teresa Pica Ciamarra, *Responsible for exhibition projects, Città della Scienza* **Italian language course**	**10 September 08** **Public activities on music and science** coordinated by Mario Campanino Responsible for Educational Activities, Città della Scienza **Public activities on food and health** coordinated by Flora Di Martino with Francesco Meo and Debora Manzoni, Città della Scienza **Workshop on the OSF plan** coordinated by Maria Teresa Pica Ciamarra with Attilio Iannitto and Antonio Tempesta, *Exhibition design, Città della Scienza* **Workshop on the OSF plan** coordinated by Maria Teresa Pica Ciamarra with Antonio Tempesta and Attilio Iannitto Atelier, Città della Scienza **Italian language course**

(continued)

Table 1 (continued)

11 September 08 **Public activities on environment** coordinated by Guglielmo Maglio with Pellegrino Covino, Riccarda D'Onofrio, Davide Petrone **Public activities on energy** coordinated by Anna Porro, with Luigi Cerri, Pellegrino Covino, Rosella Parente, Alessandra Zanazzi **Workshop on the OSF plan** coordinated by Maria Teresa Pica Ciamarra with Antonio Tempesta and Attilio Iannitto **Italian language course**	**16 September 08** **Le Nuvole: Science theatre practice** Enrico De Capoa **Case study Futuro Remoto Festival.** Diana Angela Palma, *Responsible for Communication projects, Città della Scienza* **Conferences and debate with the public** Alessandra Zanazzi and Rossella Parente, *Education unit, Città della Scienza* **Preparation of the Owerri Science Festival** by the Nigerian students **Italian language course**
12 September 08 **European awareness scenario workshop** coordinated by Luca Simeone, Città della Scienza Main facilitator: Luca Simeone, *International Relations Unit* with Alessandra Zanazzi and Alessandra Drioli, *Education and Communication projects, Città della Scienza* **Italian language course**	**17 September 08** **Preparation and presentation of the Owerri science festival** by the Nigerian students and discussion
15 September 08 **Workshop on promotion and communication** coordinated by Carlo Guardascione and Barbara Magistrelli, *Communication unit, Città della Scienza* **Workshop on visitors management and training of pilots** coordinated by Anna Porro, *Education unit*, Fiorella Iuliano, *Explainer, Città della Scienza*, Guglielmo Maglio, *Exhibitions, Città della Scienza* **Workshop on the OSF plan** coordinated by Maria Teresa Pica Ciamarra with Attilio Iannitto and Antonio Tempesta, *Exhibition design, Città della Scienza* **The chemistry bar** Elisa Rampone Chinni, Giulia Marmo Gaeta, Federico Brigida **Italian language course**	

The Science Festival in Owerri was a real party of science, with a small hands-on exhibition and science demos to stimulate the curiosity of people, especially youngsters, with few didactic materials, to be easily reproduced in the schools. It has been the occasion to evaluate through an articulate program of scientific communication the new present resources at local level: teachers, researchers and students (See Part V, From Theory to Practice).

Lesson1: Activities and Tools Overview

Guglielmo Maglio

Introduction

This chapter concerns some of the activities that *Città della Scienza* Science Centre carries out, highlighting not only the way these activities are performed and the resulting experiences, but also the positive and negative aspects to be considered once we have decided to plan and create them.

Activities

A lot has already been said about the "mission" of science centres, so I won't go any further on this point.

I would just like to underline four main goals that we, as main actors, should consider among the needs of the visitors that come to our centre:

Goal 1. Make people have fun. After all, most of our public mainly comes to the centre to have a good time. They want to have fun with science, but they also want to be surrounded by trusted people in a friendly environment. This is something that we should never forget when planning activities. An activity that we know is going to be boring right from the start should never even be considered, much less carried out! Moreover, several studies have proved that friendly and fun environments stimulate curiosity and enhance learning.

Goal 2. "Explain". After creating a good environment with good conditions, then we must provide our visitors with some basic science facts. The verb "explain" here is in quotes because we usually hear it used in academic environments. What we really want to avoid is being "scholastic". The idea here is to let people

G. Maglio (✉)
Exhibitions and Programs, Science Centre, Fondazione Idis-Città della Scienza, Naples, Italy
e-mail: maglio@cittadellascienza.it

A.-M. Bruyas and M. Riccio (eds.), *Science Centres and Science Events*,
DOI 10.1007/978-88-470-2556-1_7, © Springer-Verlag Italia 2013

free of learning by themselves. Thus, our role is to provide small bits of information which, together with the visitors' previous knowledge, helps them to create their own learning process.

Goal 3. News. Visitors expect to find experts in all scientific disciplines when they come to us. Of course this is impossible, yet nevertheless, we must be ready to deal with the latest research or "hottest" topics. Visitors often consider science centres to be their main source of reliable science information. For this reason we cannot risk losing their trust.

Goal 4. Interaction between visitors and scientists. A science centre should always be seen as a "debate" arena. Visitors should be educated to meet scientists and confront them in a "safe" place. Moreover, they should feel comfortable in debating and discussing issues with other visitors in an exchange of opinions that we must facilitate.

But *how* we can achieve these goals in our daily activities? There is no single answer. Many factors need to be considered. Here we'll just point out some of the most common problems and issues that must be taken into account when organizing an activity:

The Cultural Background of Visitors. The very first thing we need to know is the target of our activities, who is the audience and where they come from.

Our goal. Assuming we want to target a certain group of visitors, we need to keep in mind the final goal we want to achieve, the depth of information we hope to give them and the intensity of encounter that we wish to obtain.

The Time, Space and Finances Available. These factors cannot be overlooked when planning an activity! An activity will cost less if we are able to plan it in advance!

In-house Skills and Links with Experts. We should be ready to use our own in-house expertise and knowledge, but we should not hesitate to invite external experts to support us as well.

All science centres, including Città della Scienza, organise different types of activities for their visitors. The most common ones include, *exhibitions, meetings with scientists, science theatres, science shows, focus groups, science cafes* and *guided tours*. Let's look at each one in detail in order to make some observations.

Exhibitions

The most common way science centres communicate with their visitors is through exhibitions, which often perform many tasks such as posing questions, asking visitors to comment, voting or offering hints and clues. I would like to stress some of the positive and negative aspects in using an exhibition as a communication tool (Fig. 1).

Fig. 1 Exhibitions overview, Città della Scienza

One positive aspect is that an exhibition is often designed for a specific purpose. We know in advance what we want to communicate, to whom and how. Exhibitions often work well even without having someone explaining them. Thus, if we spend a lot of money in the planning and creation phases, we can also save money once the exhibition is installed. There are also ways to make an exhibition more "attractive" so that it is able to grab people's attention through colours, sounds, drawings and so on. Exhibitions can have a long life. Their message is always present and they are often transferred from one place to another, having the ability to reach a huge number of visitors often in different Countries. Exhibitions also offer different levels of information, so visitors can increase the amount and quality of information they receive by paying more attention to them or by exchanging opinions. Finally, exhibitions are an ideal platform or scenario for totally different activities, such as science shows or theatres. But I will come to this point later on.

Of course there are negative aspects too. Good exhibitions are expensive. They require a lot of knowledge and expertise and can often only provide a part of all the scientific information involved in a topic. They need to be updated (especially if they deal with new technologies) and require constant maintenance (especially if they involve hands-on exhibits). Sometimes people are too shy to approach hands-on exhibits, so they need a consultant (i.e. an explainer) we should avoid having too much text to read since we know that our visitors would prefer to "do" things, rather than "read" things.

Meetings with Scientists

This chapter includes meetings with scientists, researchers, lecturers or any other expert invited to communicate the results of a research or to give general opinions about a certain scientific topic. This is an important event for a science centre and is usually organized when a centre lacks information on a relevant scientific topic or when some specific issue is introduced and discussed with visitors. One of the positive aspects in organizing a meeting with scientists is that it allows to deal with the latest news in science and technology, even if there is no exhibition running on that topic. At the same time, it gives the opportunity to talk about "real science" (i.e. what happens in laboratories and research centres) and to try to explain it to lay people. In fact, when people have the opportunity to see and talk to a scientist outside his "typical workplace" (labs, universities, etc.) and in the friendly environment a science centre offers, they see the "human" side of a scientist and become less shy and afraid of posing questions or making comments. Furthermore, scientists need public support for their activities. This is because often activities are funded by public institutions and, consequently, by politicians who need to see that the audience supports research. It is also useful for scientists to establish links and strong relationships between centres and universities. Another positive note is that the discussion can go beyond the original issues and explore other directions, such as ethical, legal and social aspects concerning a specific field of science and technology. This enables science centres to receive direct feedback on what their visitors think about the issue, which provides enormous value!

Besides, if a famous scientist comes to the science centre to speak, the event attracts media attention as well. Newspapers are usually more interested in a single scientific event than a whole exhibition that lasts for months and is available for a longer period of time. Of course we have to consider the amount of time and money necessary for the organization of such a meeting. Famous scientists are often busy and the meeting must be organized far in advance. Yet scientists can also back out at the last minute, finding a lot of "good reasons" to avoid coming to a science centre to speak to an "ignorant" audience. So we need to have a substitute ready, even at the last minute. Another problem sometimes arises when they do come. An organizer's worst nightmare is when scientists simply don't know how to communicate with lay people or with students. In the 10 years that I have been working at Città della Scienza, a small number of the scientists I contacted proved to be able to talk and explain their opinions in a way that everyone could understand. The majority of scientists believe that ordinary people are passionate about the same things they are and what excites them is also exciting for people at large. So when I find a scientist who also knows how to communicate well, I try to do everything possible to establish long a and satisfying relationship between the scientist and the science centre. Deciding the topic of discussion and who the audience will be, can also be a problem. Normally students are the easiest audience both in terms of number and accessibility. It is easier to convince teachers to bring their students to a lecture than to convince family groups and individuals to come. Yet students tend to be more distracted and if the speaker isn't a good communicator, they can easily get bored.

Science Theatres

This problem doesn't occur, however, when we organize a science theatre in the centre. The theatre consists in a small show with scientific content performed by a young actor or an explainer with very good communication skills. It can be used as a very powerful tool to communicate scientific facts, stories and dilemmas, including those with ethical, social and legal implications. Visitors are often absorbed in the show and ask to comment and participate. Families are especially interested in these types of shows, which also reach people who are often sceptical about science as well as those who are just visiting the centre to spend time with friends or family. Since these shows work on emotions, they can easily stimulate a sense of wonder and can be adapted to different topics. They can be performed either inside the exhibition area or outside the centre. The skills of the actor or explainer can either create added value or cause the show to fail since the audience is often very critical. Actually, the number of performances per day must be limited to let the actor rest. Also, it is better to schedule science theatres only on days when a larger audience is expected. Another problem is often the huge amount of work that goes into the show. The screenplay must be written and adapted and many rehearsals are necessary before the actor learns the part. If, then, the actor should leave the centre (i.e. finds a new job), all the training sessions must be repeated. The last problem that may occur concerns the actor's performance. In fact, the actor is the real catalyst during the show rather than the story or the science fact. People, thus, may get hypnotised by the actor and forget the rest. I wrote down a comment from a teacher that came to a science theatre on Charles Darwin's life last year. He said that he couldn't hear the actress' voice due to the noise in the centre. Yet he was happy all the same because the actress was such a pretty girl that they (he and his students) just stared at her the whole time.

Science Shows

Science shows are the "daily meal" for science centres. Since the onset of these kinds of museums, the idea of organising stunning science shows followed the daily schedule of activities. Usually these shows are well-announced at the entrance of the centre and are usually repeated daily or specifically on the weekends. A science show consists in one or two explainers performing science experiments on different topics, such as physics, chemistry, nature and the environment, nutrition and so on.

More visually-impressive experiments are usually chosen and are associated to a short story or well-defined images. They are generally performed by a senior explainer who has better communication skills and typically use low cost materials so that people can try to repeat the experiments at home or at school. Watching everyday objects on the table also increases the audience's level of curiosity. For this reason it is easy to involve families and teachers (Fig. 2).

Of course these shows entail a long preparation phase as well, not to mention all the stuff which must to be cleaned or replaced afterwards. One risk involved, however, is that some scientific facts may be over-simplified in order to make them understandable.

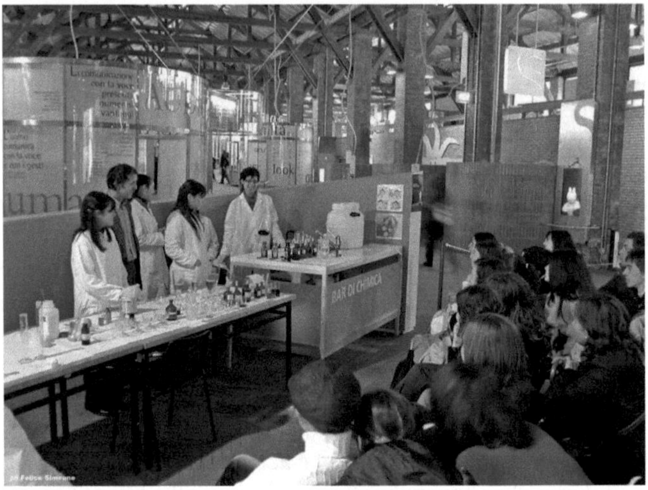

Fig. 2 Science demo on chemistry

Guided Tours

It may not seem to make sense to have guided tours in a science centre where visitors are highly encouraged to explore the exhibits on their own, but often audience is very happy when explainers bring school and family groups to discover the different corners of the centre. These kinds of guided tours are more frequent in southern Europe than in northern Europe and the US. This may be related to the cultural background of our visitors. In fact, from the very beginning, Città della Scienza realized that its visitors, when left alone, were constantly disoriented and didn't use the exhibits properly. They would even leave with the feeling that their visit wasn't worth the cost of the ticket. This was simply because they were not used to visit a museum without the help of a "resident". When we offered to escort them around, free of charge, with an explainer who could show them the most important and exciting things to experience in the centre they felt more at ease. This was because they had the opportunity to "get to know" the centre through a sort of "owner" who showed them around and because they had someone they could exchange ideas and comments with during their visit. Since then we've offered guided tours to school groups, family groups and even groups of individuals who meet for the first time within the centre.

Guided tours are one of the first ways for visitors to interact with people who work in the centre, so they need to be carried out with special attention and the explainers must be very well-trained. This is especially true when school groups come. Explainers must avoid giving "frontal" lessons and must provoke questions and comments from visitors rather than merely explaining. Moreover, explainers must be ready to understand the characters of the people they are giving the tour to and avoid having any member leading the group. They must also stimulate the more timid members of the group. The greatest risk here is having a boring tour guide or someone who is not capable of leading a discussion.

Of course there are also many positive aspects too, such as the power of a face to face contact with visitors and the possibility of establishing friendly communication. Explainers can also understand how the visitors feel towards an exhibition and whether or not it is too “cold” or too “difficult to understand”. Good explainers are able to become friends with the visitors in the first 10 minutes of a guided tour, and, thus, enhance the visitor’s experience. Happy visitors make a happy centre!

Focus Groups

This is a new kind of activity that has only recently been introduced in centres. In the last 10 years science centres have not only become places to discover science and have fun, but also places to debate the “sides effects” of science and technology. Not only the basic facts, but also the implications that S&T has on our daily lives, from the ethical, legal, and social perspectives. This has taken place all across Europe since the EU has funded many trans-national projects to interview citizens about S&T. Science centres turned out to be the best place for these interviews (or debates) because they instil trust, give voice to different poles of a science debate and because they help visitors create their own opinions. These discussions have taken place all over Europe and special tools have been developed to help visitors debate scientific topics, even when they lack the necessary knowledge to do so. Focus groups, instead, usually consist in organized meetings between a facilitator and five to ten people of different age, sex, cultural and social backgrounds. Once they are in the centre they watch a short movie, listen to a story, visit an exhibition or interview a scientist for a short time. This event is called the “starter event”. After that, the facilitator starts the discussion and helps participants to express their opinions and to discuss what they have seen, listened to and visited with the other members of the groups. Sometimes this discussion is tape-recorded and successively analysed by sociologists and other scientists in order to understand the way the discussion has developed.

The positive thing is that people are very happy to debate and understand the opinions of others. They realize that, with the proper help, they are able to discuss topics that may have been completely unknown to them before, making them feel that they have a role in this democratic process. It is very important to know what our visitors think about topics related to science and daily life. This helps us to build better exhibitions, programs and activities and also emphasizes the social role of museums and science centres, helping people form their opinions and allow then to vote in democratic processes. Sometimes, however, the facilitator isn’t able to guide the discussion in a fair and unbiased way. Focus groups are also a bit expensive and, worst of all, it is very difficult to convince people to take part in an hour long public discussion when there are so many things to do and see all around them. It usually helps to offer them food and drinks!

Lesson 2: The Role of Exhibitions in Science Museum and Science Centre

Anna Porro and Luigi Cerri

A new type of museum based on newly created materials for learning, exploration and discovery was created in the second half of the twentieth century. Science centres find their roots in the working technology exhibits displayed in the Deutsches Museum in Munich in the 1920s. The Evolution of Eindhoven, the Exploratorium in San Francisco and the Ontario Science Centre sprung up in the 1950s–1960s and represent the first real examples of science centres as we know today, presenting interactive exhibits throughout the whole world. The collections in these types of museums require a specialised type of workshop space.

As it is well-known, visitors in these museums are not only invited to look but also to take part in. Thus, exhibitions must be designed for demonstrations and hands-on activities both in the galleries and in adjacent workshops. Creating a good hands-on exhibition requires passion, intuition, knowledge and skill. It is a creative activity that involves curators, planners, designers, technicians, scientists, educators and, if possible, samples of visitors. The challenge is to create participatory and interactive exhibits that are easy to use and which encourage people to experiment and explore.

Frank Oppenheimer, the founder of the Exploratorium in San Francisco, says that it is important to plan for variation as much as possible because only a limited amount of understanding comes from watching something behave. One must also watch what happens when one changes the parameters that alter behaviour.

Participation and interaction personalise the exhibition environment. Interactivity attracts, engages and wakes people up. Interactivity is about being "reciprocal". The visitor acts upon the exhibit and the exhibit does something that acts upon the visitor. The most important aspect of an interactive exhibit is the feedback people

A. Porro (✉) • L. Cerri
Education and Science Communication Projects, Fondazione Idis-Città della Scienza, Naples, Italy
e-mail: porro@cittadellascienza.it; cerri@cittadellascienza.it

A.-M. Bruyas and M. Riccio (eds.), *Science Centres and Science Events*,
DOI 10.1007/978-88-470-2556-1_8, © Springer-Verlag Italia 2013

Fig. 1 Exhibit on reflected colours

receive. If the results of an activity are too subtle, or if the payoff requires reading paragraphs of explanatory texts, the concept may be lost or misunderstood and may cause frustration. When feedback is insufficient, people may walk away from an activity at the half-way point, leaving an exhibit with inaccurate information and incorrect assumptions.

Make sure the exhibit works. People are wonderfully unpredictable and they interact in an unlimited variety of ways. It is imperative that exhibits have a simple design, clear and precise instructions and understandable results. People will not spend much time trying to figure out how to use or operate something. Interactive exhibits require a maintenance staff to keep them going, no matter how well they are designed and built (Fig. 1).

Time is an other important factor in any exhibition experience and people today never seem to have enough of it. Consequently, planners have to design for clarity and brevity. Visitors have limited time to absorb information, so interpretative graphics should be clear, concise, understandable and short. People usually spend only few seconds reading a label, which must convey essential information in that time. It might be assumed that the creation of labels is a relatively minor function in the grand scheme of exhibition development. Several planners are often convinced that people do not read labels, but studies have shown that people want information about the exhibitions they see and a lack of labels produces lower interest levels. Exhibitions are primarily *environments for communicating ideas* and labels represent the conversational voice of an exhibition. The *challenge* is to produce exhibit texts that people read and easily comprehend. For an effective communication, exhibitions must speak in a familiar language and tell interesting stories.

Fig. 2 Exhibit on colours and lights

Moreover, exhibit planners must know who their message is directed to, what their interests are and what common ground or prior knowledge exist. If possible, exhibits must be able to speak the language of their visitors and be *clear and simple*, but not *simplistic* (Fig. 2).

Lesson 3: Design Process for the Organization of Exhibition Areas

Maria Teresa Pica Ciamarra

Introduction

An exhibition is a special learning tool that offers a narrative approach to the understanding of complex issues. It is also a place where both emotions and reason can live together harmoniously.

Although the main element of an effective scientific exhibition is the scientific theme, the way in which it is designed and realised can deeply influence its overall effectiveness.

The exhibition design requires creative and organizational skills in favour of exhibit development, scenography and project management, as well as expert know-how of financing, copyright-management and the international transfer and lending of exhibition objects. In addition, professional multimedia-based presentation techniques and experience in virtual exhibitions are needed.

In general, the exhibition-creation process can be organized into 3 main steps: (1) Planning and Design, (2) Realization & Installation, (3) Maintenance & Implementation.

Planning and Design

The general design of an exhibition should be determined by carefully analyzing the purpose of each exhibit and the expected audience, as well as the amount of funding available. The most appropriate style for a scientific exhibition will change depending on the type of scientific topic to be presented, the equipment and the budget. This is especially true with permanent or touring exhibitions. An exhibition

M.T. Pica Ciamarra (✉)
Exhibition Development, Fondazione Idis-Città della Scienza, Naples, Italy
e-mail: mt@cittadellascienza.it

A.-M. Bruyas and M. Riccio (eds.), *Science Centres and Science Events*,
DOI 10.1007/978-88-470-2556-1_9, © Springer-Verlag Italia 2013

Fig. 1 Meet Math exhibition

is, in any case, a method of communication that integrates different kinds of media. Moreover, it can be considered a "body & soul" system.

The *body* corresponds to the elements physically present in an exhibition, mainly the scenography (the general context) and the exhibits. There are many different ways to convey complex scientific information in an exhibit. This also depends on the public the exhibit is targeted at. Exhibits may be static, as most posters are (i.e. didactical graphic panels, deepening information panels), or interactive. Choosing the most appropriate type of exhibit to effectively convey the specific data is the first and most important step in creating a valuable scientific exhibit. Exhibits can be broken up into didactical and graphic panels, hands-on exhibits, multimedia exhibits (audio & video products), artistic artefacts and/or real objects and internet-based exhibits (that can be integrated to bring different audiences together). Experiences, information, visual and interactive elements, eye-catching and attractive artwork are all used to better explain scientific topics, as well as to stimulate interest, scientific curiosity and create enjoyment. In addition, an exhibition has to create a context where people can exchange feelings and share opinions about the scientific topics offered (Fig. 1).

The *Soul* of an exhibition is its general atmosphere, which must initiate from the specific theme, message and thinking geared towards a specific target audience. Activities linked to the "body" of the exhibition, such as events, educational tools, didactic activities, theatre shows, etc. also contribute to the spirit of the exhibition, making the role of pilots and communicators essential for a successful outcome. Thus, the exhibition's mission (integration and dialogue between *body* and *soul*) is therefore to provide the answer to its objectives and specific aims, as well as to its scientific content, concept and target audience.

Aims. The first goal must be to create a pleasant general atmosphere that "talks" to visitors and invites them to enter the exhibition and approach the single exhibits.

Scientific content. The exhibition must be directed to the public, beginning with the concept of the exhibition and from the selection of science items and sub-items to include in it.

Fig. 2 NanoDialogue exhibition

Target audience. The final beneficiaries of science exhibitions are normally school students of different ages, families and the public at large. The whole exhibition and each individual exhibit must be addressed to each specific audience. For this reason it is imperative to know and understand the audience's cultural background and social context in order to choose the specific communication tools for that particular final user (especially when intended for children).

Management. The exhibition should also support educational operators in their work by creating appropriate areas, exhibits and inspirational elements for the didactic activities that will be carried out (Fig. 2).

Organisation of Work

In order to start the planning of the exhibition, it is important to set up work-groups, to organise the work into activities/phases and to split up the responsibilities among each of them. The delivery of the work must be structured through a time schedule related to the different activities and phases of the project. The time-table must be checked continuously as work progresses and, if necessary, updated (Fig. 3). It must always be shared with the entire work group as well. For example: finally, the costs from the project phase to the production, installation and testing phase must be analysed, taking into consideration technical and other specific needs as well (Fig. 4).

General Planning and Scenography

The general layout should include the special "nature" and atmosphere of the exhibition, which, as previously mentioned, should be correlated to the main

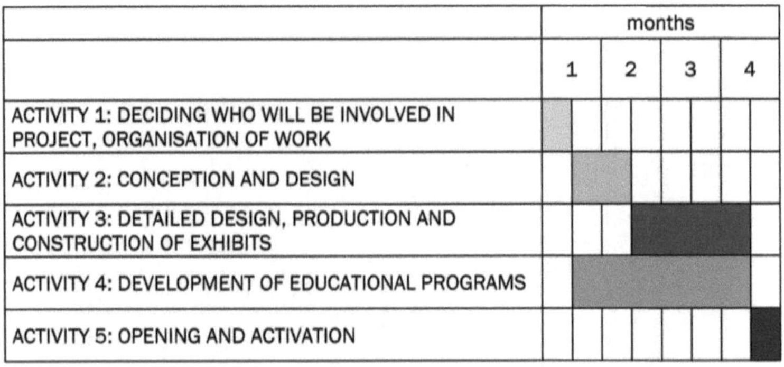

Fig. 3 Example of exhibition development timetable

scientific message as much as possible. General planning includes the choice of principal materials, technology and colours to employ for the realization of original exhibits. The logo of the exhibition must be projected too. During this phase, it is crucial to consider the impact of the exhibition's design and space arrangement with regards to didactic aspects and activities.

Furthermore, the exhibition is strictly linked to its venue, so it is necessary to confirm important information about the place, such as:

- Analysis of the internal and external areas in order to create a comfortable atmosphere and guide the public through the exhibition;
- Checking all technical characteristics (e.g. lights, electrical connection and plugs, internet, accessibility to the exhibition for the disabled, etc.)
- List of the possible solutions of problems (e.g. accessibility of exhibits into the inner area for the mounting phase, electrical and data connection lines, etc.).

Planning the Individual Elements of the Exhibition

Make a list of all the elements that make up the exhibition (exhibits, didactic and graphic panels, furniture, technical equipment, multimedia products, audio-video, etc.) and a list of the industrial elements that need to be bought (technical equipment, furniture, etc.). A first draft of the didactic panels must also be created offering clear and simple information to communicate an idea. Then more complex panels must be created to explain more complicated information.

Multimedia exhibits, immersive displays and workspaces (i.e. new technology-based exhibits) generally arouse the public's interest. The design of an exhibition is increasingly more shaped by advancements in both computer and video technologies. Recent progress in the development of high-definition displays offers the possibility of reusable interactive displays. Interactive technology will play a significant role in the future as it becomes more and more employed. Many exhibits

	Unit	# of units	Unit rate (in EUR)	Costs (in EUR)
Equipment and supplies				
Exhibition materials				
General exhibition scenery	per sqm	400	€ 15	€ 6,000
Hands-on exhibits	per exhibit	20	€ 2,500	€ 50,000
Multimedia products	per product	4	€ 200	€ 800
Panels	per panel	40	€ 40	€ 1,600
Repairs and modifications during the Festival	per week	4	€ 100	€ 400
Furniture, computer equipment				
computers	per comp	8	€ 700	€ 5,600
TV screen integrated in the exhibition	per screen	4	€ 350	€ 1,400
flat screens	per screen	2	€ 400	€ 800
touch screens	per screen	-	550	-
plasma screens	per screen	2	€ 700	€ 1,400
Video projector	per projector	4	€ 350	€ 1,400
Spare parts/equipment for machines, tools	per tool	30	€ 100	€ 3,000
Subtotal Equipment and supplies				**€ 72,400**
Publications				
Posters & leaflets of the exhibition	per copy	400	€ 30	€ 12,000
Didactical materials	per copy	1000	€ 4	€ 4,000
Exhibit transport, insurances				
Exhibit / other transport	per mc	400	€ 25	€ 10,000
Subtotal Other costs				**€ 26,000**
Total cost				**€ 98,400**

Fig. 4 Example of exhibition budget

already encourage active participation on behalf of the visitors through the use of three-dimensional pieces of glass and movable partitions. Computer and touch-screen sensors will take these developments to an even higher level, which will blur the distinction between static and interactive exhibitions. Fully active scientific exhibits of the near future will permit full-motion videos, interactive choice of data, as well as the ability to both actively and passively supply feedback to authors and designers. These technologies usually require a high budget, both for the purchase of relevant technology (hardware) and for the creation of original products

Fig. 5 BIT virtual puppet

to present (software), as well as for the connection systems needed (i.e. synchronous tools) (Fig. 5).

Easier and cheaper exhibits, however, have excellent response rate too, in particular participatory exhibits where visitors can share experiences and exchange opinions. "Debate stimulation" on scientific topics is definitely a great achievement in this direction and, in fact, developing stimulating exhibits must become a priority.

The detailed project and design of an exhibition will include all specifications for the creation of the different elements:

- General layout of the exhibition – including specifications for electricity and data lines, lights, logistic and specific needs.
- Design of single exhibits – including information about materials, electrical/mechanical parts needed, colours, finishing, etc.
- List of elements to be bought – such as furniture, technical equipment, etc.
- Graphic design for didactic panels, signals, scenography – including specification of the kind of print process, backing materials, fixing systems, etc.
- Story-board for the creation of original multimedia products.

- Final check of general and individual costs – including shipping, insurance policies, etc.

Creation and Installation

During this phase, the designer is responsible for the direction of work and stage management, both during the creation of exhibits and the mounting of the exhibition. Many problems can arise throughout this period. The designer must be prepared to make changes to the project, if necessary, in real time and solve problems without hesitation. One may have to revert back to some quick brainstorming for possible solutions, analysis and decision techniques in order to respect the time table. Some changes may become necessary right before the exhibition opens as there might be problems with the working order of an exhibit or with text information, etc. Once the different exhibits are installed, each of them must be set up and tested. A sample group of visitors could test each exhibit and give feedback about the aesthetics and functionality of the exhibit as well as its didactic aspects.

Maintenance and Implementation

In order to guarantee the good maintenance of the exhibition, the designer must communicate maintenance information to the staff. A technical "manual" could be created for such purposes. A visitor satisfaction analysis is also imperative to evaluate the impact of the exhibition on the different target audiences that participate and the general atmosphere of the exhibition on the whole as well as the single exhibits.

Effective exhibitions require careful planning, the use of existing equipment, personnel and hard work. Each type of exhibit has advantages and disadvantages and should be selected on the basis of content, target audience, type of scientific topic as well as the resources available, including funding. Analysing the main intention of the exhibition, the expected audience and information, it is possible to create an exhibition that is attractive and effective for all.

Finally, from the very first stages of its development, an exhibition must continuously be open to change in order to better meet visitor expectations with regards to the dissemination of science for educational purposes, the general appeal of the exhibition, the working order of the exhibits and scientific updates.

Lesson 4: Educational Activities

Mario Campanino

Introduction

This chapter tells how experts and educators of Città della Scienza's science centre have involved classes of students in the planning and testing of an innovative method in teaching and learning science. This method is an important tool that educational institutions can utilize in the development of increasingly dynamic and different educational paths.

Activities

During Laboratories for Science Education (LES) activities, experimental didactic experiences take place in non-formal learning environments. Here every classroom element (from furniture to equipment, to the language used by the instructors and facilitators of the experience) is designed to encourage an active and collaborative learning environment. Student participants range in age from 8 to 18 with a focus given to the specific participant's age when developing relevant activities. There are approximately 150 different educational opportunities presented in a diverse catalogue of topics that include disciplines such as physics, mathematics, earth and life sciences, and even art and musical science (Fig. 1).

M. Campanino (✉)
Education and Science Communication projects, Fondazione Idis-Città della Scienza, Naples, Italy
e-mail: campanino@cittadellascienza.it

A.-M. Bruyas and M. Riccio (eds.), *Science Centres and Science Events*,
DOI 10.1007/978-88-470-2556-1_10, © Springer-Verlag Italia 2013

Fig. 1 Teachers training on LES activities

Learning and Knowledge

Città della Scienza's exploration into teaching and learning science began as most subject research does, through the desire to focus on the development of new concepts and to find answers to questions without definitive answers. In this instance, the first question analysed was: what is learning? Subsequently we analysed: what is knowledge?

A Berkeley University study reveals that the quantity of written information produced by humans between the year 2001 and 2004 is equivalent to the quantity produced from 1970 to 2000. Additionally, this quantity is equivalent to all written text produced since the invention of writing to 1970. This exponentially increasing trend is further illustrated during the years 2005–2006, when mankind again produced the same quantity of information as the aforementioned periods, but within a shorter timeframe.

Additional questions arise in how human beings can manage such large quantities of information. Also, how can instructors navigate themselves and their pupils (in an educational context) through this massive maze of knowledge? The effective transfer of such information needs the establishment of an alternative to the traditional teaching method. This alternative methodology should stress the appropriate manner of learning and acquiring knowledge while respecting that a student, once cognizant of an appropriate learning strategy, will be able to properly and independently continue their pursuit of knowledge through effective use of all readily available information.

Non Formal Methods and Constructivism

Constructivism is a pedagogical theory that proposes a new way to teach and to learn aimed at the assumptions previously represented, where the teaching/learning process is conceived in a non-traditional manner beginning with the concept of 'learning environment.' A constructivist learning environment must be meaningful, active, collaborative, conversational, reflective, contextualized, intentional, and constructive. As Dave Jonassen, a researcher and professor in the fields of educational psychology and learning technologies, has noted, a constructivist environment must:

- Give emphasis to knowledge construction, not to knowledge reproduction.
- Avoid extreme simplifications in representing the complexity of real situations.
- Present true tasks (to contextualize rather than to abstract).
- Offer learning environments coming from the real world, case-based, rather than predetermined instructive sequences.
- Offer multiple representations of reality.
- Encourage reflection and reasoning.
- Allow knowledge construction not necessarily linked to specific content.
- Encourage the cooperative construction of knowledge, through collaboration with others.

As such, the non-formal context may be defined as an environment that expresses educational values but that is not subjected to external control. Non-formal is a characteristic linked to the learning environment, the methods, and the dynamics of learning. It finds its philosophical basis from what Bruno Munari declared many times in his writings:

- All people can complicate, few people are able to simplify.
- To simplify you have to know how to remove; and this means you have to know what to remove.
- It is more difficult to simplify than to complicate.
- It is more difficult to remove than to add.

Society typically has defined teaching as 'explaining' while equating learning to 'studying.' Alternatively, constructivism proposes that people begin to think of teaching as 'facilitating' and 'participating' in educational activities and learning as 'doing,' 'talking,' 'imagining,' 'planning,' 'producing,' and 'evaluating.' In fact, constructivism tells us that to know and understand reality means to successfully associate facts and meanings. World experiences such as thoughts, emotions, and sensations act as signs, and meanings are assigned to these signs during the process of learning. Thoughts relate to language and to formalized knowledge are part of a human being's rational side (brain); a human's non-rational side relates to its emotions and feelings (heart); sensations relate to sensory organs (body). All three types of human experiences are equally important in terms of the development of human knowledge; as such, educational learning activities should be more than just reading a text book. Educational activities such as student participation in experiments, demonstrations, games research, verbal communication, and performances should be incorporated into the curriculum of both formal learning (e.g. school) and other learning (e.g. museums, science centres, science events) institutions.

Evaluation

One of the main issues in non-formal educational activities relates to the proper evaluation of student progress. The problem arises in determining 'what' and 'how' to evaluate. In the non-formal educational process, there are few clear quantitative indicators of the success of both teaching and learning process. As such, the evaluation process must begin by defining the goals of the educative initiative. For example, what do we want to teach with non-formal practices? One constructivist-based answer would be that with the non-formal education we want to improve the quality of the knowledge process by changing the 'learning behaviour' of pupils. How can this be measured in order to evaluate progress? The answer is by examining how the student behaves during the educational activities, how they organize their knowledge, and what are the student's personal thoughts on the experience. Therefore, the assessment tools needed to collect this type of data from the students would not be considered typical in a traditional school environment.

First, to evaluate how a student behaves during the activity, the following tools could be used:

- Audio and video recordings of all the activities;
- Interviews with the facilitators of the activities;
- Interviews with the teachers which are present during the activities.

Second, to evaluate how they organize the knowledge, the following tools could be used:

- Conceptual maps;
- Questionnaires;
- Drawings (for kindergarten and primary school level students).

Lastly, to evaluate what are the student's thoughts on the experience, the following tools could be used:

- Reports about the learning experience;
- Letters to a friend (a real or imaginary friend, to whom students have to communicate the purpose of the learning experience);
- Phone calls to a friend (the same of the letters, but more linked to emotional aspects);
- Interviews with the students.

As it can be easily imagined, the evaluation process is a very difficult task. In my opinion, it is one of the primary reasons that have prevented the inclusion of non-formal practices in the teaching and learning methodology applied in today's schools and other educational institutions. Nevertheless, it is currently impossible to think to an effective science education without referring to this method or even more importantly, educational philosophy. In fact, this is the only way to help the pupils of today to face the complexity of the modern world.

Lesson 5: Conferences and Debates

Rossella Parente and Alessandra Zanazzi

Introduction

One of the most important ways to interact with the public in a science centre is through lectures, conferences and debates that involve the public on one hand and scientists, researchers, experts and decision makers on the other. These events can complement formal learning by supplying knowledge and in-depth investigation as well as by providing citizens with a forum to address critical science and social issues. These speeches, public talks can be organised in many different ways and effect/cause different levels of involvement and participation of the public. In any case, the lecture/debate should be carefully organised in order for the speaker to communicate ideas effectively and for the audience to understand, learn, critically analyse and question the speaker.

Planning

When planning an effective event, organisers should ask themselves some key questions:

- WHO? Who are we addressing? What is the age and type of audience we are dealing with? Who is the speaker we want to involve (depends on both the topic of the event and personal attitudes)?
- WHAT? What do we want to communicate? Hot topics arising from current social debates? School subjects? Cutting-edge science issues?

R. Parente (✉) • A. Zanazzi
Education and Science Communication Projects, Fondazione Idis-Città della Scienza, Naples, Italy
e-mail: parente@cittadellascienza.it

A.-M. Bruyas and M. Riccio (eds.), *Science Centres and Science Events*,
DOI 10.1007/978-88-470-2556-1_11,

- HOW? How do we want to interact with the audience (i.e. lectures, round tables, café scientifiques)?
- WHEN? Formulate a schedule.
- EVALUATION. What do we evaluate? What parameters do we use to make an evaluation and with what tools? How do we select target groups?

The first 3 points are obviously the most decisive, but they do not have to be put in this order. There is no recipe. You may decide to organise an event for school students regarding a key topic specific to the year in reference. This is your starting point. Now that the topic (the *what*) and the audience (the *who*) have been settled, you can proceed finding the right speaker and deciding the *how*. You may want to have a Nobel laureate visit and let the public meet him/her. Or you may want to have citizens participate in a discussion on a 'hot topic' that is being widely debated in the media and you may need to involve different speakers with opposing viewpoints, etc.

Who?

Who are the people you wish to "talk to"? How old are they? What is their living environment? What is their family environment? What is the learning environment (formal or informal)? Are you addressing the general public or schools? What is their education and literacy level? This is not at all a trivial matter. It is crucial to know the public you want to involve and that your speaker is informed about who his/her audience is in order to use the correct language. Sometimes speakers are not able to adapt their language to a certain audience. Thus, you must be careful of avoiding experts who do not know how to simplify complex concepts for children and families or, conversely, someone who simplifies too much for an educated audience or who may be too far from the world of listeners (e.g. knowing how to speak to teenagers). You must also consider the specific categories of people that may comprise your selected audience (i.e. teachers, students with special needs, etc.).

What?

The subject is of vital importance. A good conference normally provides an answer to a presumed need and the best topics are those that are not already addressed in a satisfactory way by the formal curriculum within schools or those that arouse curiosity, controversy, or are just simply interesting to debate about. There are two possible approaches to chose a topic: a bottom-up or a top-down approach. In the first case, you have been asked to answer some specific requests from your target audience (e.g. school teachers would like some in-depth investigation on a certain subject or a request has emerged from a questionnaire, survey, etc.). In the top-down approach, as example you may also act as an intermediary between the participants and other institutions (i.e. Universities, Research Centres, Ministries, NGOs, Churches, Activist groups, etc.) who want to inform the public of new discoveries they have made and the importance of their research, to communicate decisions, behaviours, carry out an awareness campaign, etc.

How?

In any case, the language of communication plays a role just as fundamental as of *how* you want to communicate, or rather, the kind of event you wish to propose and which is mainly associated with the target audience you have identified.

Possible events include: lectures, interactive lectures (Fig. 1), conferences, round tables, café scientifiques, science shows, science theatres, and the list goes on. Let's analyse some of these in a greater detail.

Lectures and Conferences. One or more events with one or more speakers that present specific researches, the current situation of a question, basic science, etc. These types of events do not foresee interaction with the public except for the questions at the end of the talk. Given the low level of interactivity, this type of event is easy to manage, provided that you inform the speakers about the subject, the target audience and other lectures and events being held at the same conference cycle (if this is the case). The event shouldn't last more than 2 hours, otherwise you risk losing your audience. And you must always leave time for questions! A good way to stimulate participation is by distributing some materials before the conference (i.e. to a school class). Also, take-home material is another useful tool to ensure some follow-up. Some tips include: having more than one speaker and changing the tone. So setting up a dialogue or even a debate on opposing (science-based) theories also keeps attention levels higher. It is easy to provide up-to-date information and organise a meeting between scientists, researchers and lay people, but you have to carefully plan and advertise the event if you don't want to end up with an empty conference room! You don't necessarily need a 'famous' speaker to attract participants, even though it does help! It is important your speaker uses a proper language and is able to convey enthusiasm, passion and interest during the lecture. Someone that lacks conviction will never be able to enthuse and persuade others!

In order to increase participation you can also contaminate the "classical" lecture by introducing some science show elements and having a "show-lecture". For example, during the talk the speaker can perform some experiments, especially visually-impressive ones (e.g. those that produce smoke, lights, etc.). This is helpful in both explaining science principles and in creating involvement. If a speaker knows how to 'perform', an audience of adults has no problem in playing and interacting.

Another way to spark interest is to link the conference to the exhibition. For example, a guided tour of an exhibit, model, mock-up etc. could follow the lecture.

Videoconferences. Videoconferences are another possibility. They function the same as a lecture would, but have the advantage that your speaker does not need to be physically present. You will need, however, a projection screen, an Internet connection, a webcam, a microphone and loudspeakers. Once you obtain the technical equipment, it is easy to organise and is especially fun for youngsters. It also allows "access" to other infrastructures such as talking to experts located in a faraway place. Some inconveniences may include technical difficulties or translation problems (as with foreign speakers). Lastly, interaction (questions and debate) is much more difficult on the whole.

Round tables. When a moderator/introducer and some speakers discuss researches, science and social issues, etc. It does not necessarily have to follow a prepared presentation. No interaction with the public is foreseen, if not at the end for questions. It is imperative that the speakers know what the subject matter is, what topics are covered by other speakers, who the audience is, the materials distributed (if any) and what the other conferences and events running at the same time. The difficulty here is in finding a good moderator who is impartial, keeps time and, most of all, creates stimuli for the debate. Once again, the event must not last more than 2 hours, otherwise you risk losing your audience and you don't leave time for questions! The advantage of a round table is that it allows you to compose different theories, viewpoints and experiences as well as demonstrate a certain subject through different perspectives (i.e. you can invite a biologist, a doctor and an expert on ethical/legal issues etc. to discuss new frontiers of nanomedicine). This gives a more complete panoramic view on the topic and on its complexity. Choosing the speakers and the moderator is complicated here as well (language, enthusiasm, passion, respect, etc.).

Café scientifiques. Forums used to discuss interesting scientific topics in an informal and accessible way (whereas a lecture, conference or a round table are more formal approaches). In fact, this approach involves having a speaker (and a moderator if this is the case) who discusses about scientific subjects in a bar, pub or bookshop while everyone enjoys a cup of coffee or tea, a glass of wine or a drink. There is immediate interaction and debate. The speaker is normally a scientist or a science writer. The event normally starts with a brief introduction of the speaker and a short talk, leaving ample time to converse and discuss with the participants. Everyone can ask questions and there is usually a big turnout. The environment plays a key role because café scientifiques aim at de-mystifying scientific research and bringing it closer to home for the general public. Thus you must pay special attention to choosing the right place. Café scientifiques are not normally held inside places devoted to science or to science communication and education. They may be held in downtown bars, restaurants, theatres, etc. Once again the speaker is fundamental because he/she has to have an informal attitude and must be able to adapt his speech to the audience and follow the debate in any direction it may go, depending on the curiosity of the participants and the questions they ask.

Fig. 1 Open lecture in the Science centre

When?

Plan the schedule as much in advance as possible. This includes:

- *Preparation*: who (speakers to involve, target audience), what, how and where (a good location). You must also provide the needed resources. Speakers that use PowerPoint presentations, movies, images, slides, overhead projections, objects, etc. will be more likely to involve their audiences.
- *Implementation*: write the programme and abstracts, call/write to speakers and participants, "advertising". "Advertising" is fundamental. You can use local radioas, newspapers, websites, bill-sticking, posting, mailing lists (both email and surface mail if possible), group advertising (churches, interest groups, schools, teachers, etc.)
- *Producing Documentation to be distributed*: pictures, recordings, PowerPoint presentations, materials, etc.
- *Evaluation*: questionnaires, conceptual maps, interviews and data analysis, auto-evaluation (log registering of what, when, why, expected and obtained results).

Evaluation

Evaluation must cover:

- The general satisfaction level of the target group;
- The motivational impact of the activities (changes in perception of the relevance of the topic and curiosity towards it);
- The educational impact of the activities (understanding, knowledge improvement).

Possible tools could include questionnaires with open and closed answers and space for free comments, interviews and conceptual maps (pre/post) that can be used in order to understand changes in the lexicon of the target audience.

Lesson 6: Science Festival

Diana Angela Palma

Introduction

"Futuro Remoto. A Journey Between Science and Science Fiction" is a multimedia festival to spread scientific and technological culture held annually in Naples. From its first edition in 1987, the festival has aimed at facilitating the understanding of some notable contemporary scientific questions by focusing on the interdisciplinary characters of several topics and by highlighting the social consequences of scientific and technological innovation on our lifestyle. Participants mainly include students, young people and ordinary people who are interested in scientific and technological innovation.

Futuro Remoto is based on a specific theme that changes every year (i.e. biology, energy, environment, robotics, etc.). It includes several events such as the main scientific and interactive exhibition, a series of lectures, an expo of scientific and technological innovations, scientific and artistic labs, video projections, performances and activities for children. The main exhibition covers a specific theme for which a series of lectures are held regarding scientific issues raised by that same theme as well as amazing activities.

Choosing the Main Theme

The main theme is chosen through "brainstorming" techniques and is geared towards the specific target audience of the festival. Some parameters that help in choosing a theme are:

D.A. Palma (✉)
Futuro Remoto Science Festival, Fondazione Idis-Città della Scienza, Naples, Italy
e-mail: palma@cittadellascienza.it

A.-M. Bruyas and M. Riccio (eds.), *Science Centres and Science Events*,
DOI 10.1007/978-88-470-2556-1_12, © Springer-Verlag Italia 2013

Fig. 1 Futuro Remoto 2002 'Questions about our future'

- *Appeal.* A theme chosen in order to attract a large audience for promoting our Science Centre (Fig. 1).
- *Topicalness.* A theme chosen for its relevance at that present moment (i.e. the Olympic games or, in 2012, the International Astronautical Congress).
- *International Awareness.* A theme chosen in order to respond to common international needs that are particularly requested by international community (i.e. Global Changes, which have the advantage of receiving funding specifically set aside by the European Commission).

Teamwork

Experts are put into groups, or teams, that are led by a project manager. Each group is responsible for one of the assigned tasks:

- Developing Festival Content (exhibition, labs, lectures, activities for children).
- Finding Sponsors.
- Advertising.
- Marketing.
- Exhibition Layout.
- Maintenance.

The project manager oversees the project as a whole, from the first idea to the end of the festival. For the *Futuro Remoto* Festival, the project manager is also in charge of managing the budget, drafting the storyboard, defining times and places, establishing scientific committees (experts who develop and check contents),

Fig. 2 Futuro Remoto 2009 'Robot'

searching for possible partners (universities, research centres, companies), requesting patronage from institutional and scientific authorities (in order to gain prestige for the event) and coordinating different activities (sponsors, communication, development of content, etc.).

Developing Content

The scientific exhibition is the main event of the *Futuro Remoto* Festival. There are three options for setting up a scientific exhibition: create a brand new exhibition (which requires lengthy development time); rent an existing travelling exhibition from another museum (short development time as the exhibition only needs to be adjusted to our logistic needs, yet provides a costly solution); rent a collection of elements from a museum and develop the exhibition based on these elements, including additional materials taken from other sources as well.

Once the exhibition has been decided upon, the agreement with the Institute lending the exhibits must be defined. The agreement must specify the period of the exhibition, rental expenses, the insurance value of all single elements on display, the name of the shipping company, the name of the person responsible for overseeing exhibition layout (plus the number of technicians and days required) and a detailed list of the elements rented (Fig. 2).

The exhibition may also be enriched by adding other elements and exhibits that you might ask to other museums, universities, research centres or companies. These extra elements may be offered in exchange for visibility and free advertising at the festival. The science centre, however, is responsible for transportation, insurance and layout expenses in these cases.

Fig. 3 Futuro Remoto 2010 'Form prehistory to the cities of tomorrow'

The exhibition is like a story that must then be divided into chapters, so that each chapter becomes an expositive section. Each section could include a big or an important exhibit, some other small exhibits, as well as some multimedia and text panels. The topic of each section could then be developed through all these elements. It is also important to remember to keep text panels short and clear (approx. 10 lines each)!

Lectures

The project group must also identify possible topics of scientific interest within the main theme of the festival so as to widen its breadth. Lectures propose several new topics each day, allowing the media to renew interest in the festival almost every day.

Layout of Exhibition

Once we have developed the content and chosen all the elements for the exhibition, we can begin planning its layout. The design group is in charge of this task, but the project manager has to oversee this stage of the project as well, in order to make sure the concept, contents and expositive outcomes all together.

It is imperative for the project manager to consider *the number and kind of visitors foreseen in the exhibition area each day*. It is particularly important to know how many students can all stay together in the exhibition area. The visitors' path must also be defined at this stage in order to avoid trouble after the exhibition opens (Fig. 3).

Advertising, Sponsors and Marketing

These tasks are dealt with the advertising and AS&M group (see Carlo Guardascione's chapter). At this stage the project manager works together with the advertising group and the designers to develop the graphic image for the festival.

Communication Materials

It is also important to create communication materials:

- Web-based materials are capable of raising the interest level of a large audience and must have great appeal in order to convince people to come to the festival.
- Communication materials for the press must include main features, news and the most interesting aspects of the festival in order to persuade a reporter to give more importance to our festival instead of other events. In the example of our festival, a persuasion factor may be that this is the first time an original dinosaur exhibition from China makes its way to Italy. Secondly, some of them are feathered fossils (meaning that they had wings like birds), which have brought about a revolution in studies in palaeontology. Thirdly, prominent scientists have been invited as lecturers.
- Materials distributed in schools, however, must give an overview of the entire festival, focusing on matters that teachers would consider interesting for their students in order to persuade them to bring their students too.
- Other materials have to be developed for the scientific explainers inside the museum. These materials must put forward the contents of the exhibition with detailed descriptions about every single element on display.

Technical Issues

Technical issues will no doubt arise as work on the project progresses. It is important to handle these problems immediately in order to avoid further problems. A shipping company must be found for the transportation of the exhibition elements. Call several companies and choose the lowest estimate to save money. Do not forget to draw up a formal agreement. Insurance coverage will be needed for valuable elements for the entire duration of the exhibition, including shipping periods.

Maintenance

Do not relax the day after the opening! Work must go on! First of all, it is important to pass all useful information about the festival to information and operation staff and maintenance. This information includes opening and closing times, the path that the visitors must take through the exhibition, the daily lectures and events (labs, planetarium shows, activities for kids, performances) schedule and location, number and kind of students expected each day and the list of people involved in each of the daily activities. It is also important to make sure there are people in charge of logistics and computer services throughout the entire day, from opening to closing.

Increasing Numbers with *Futuro Remoto*!

When *Futuro Remoto* opens, we generally have a dramatic increase of visitors, requiring more staff at the ticket office, more security and more scientific explainers. Other services related to the museum, such as the shop and café, must be prepared to handle greater crowds as well.

Finally, do not forget to thank your partners for taking part to the festival! You can insert a word of thanks or thank you card into the brochure and you can send cards to thank the people involved in the project at the end of the festival.

Lesson 7: Communication and Media Plan

Carlo Guardascione

Introduction

Communication plays a strategic role in the success of an event. Communication is not just a process whereby information is transmitted with regards to a scientific event. Communication represents the power to create images and symbols in order to generate great emotions and suggestions.

When an event is carried out, communication plays a *strategic* role rather than a mere *organizational* one.

Do not communicate an event is like have not the event at all! The communication plan must be worked on from the very beginning. The following phases must be carefully worked through when designing a communication plan:

Choosing a Theme

The chosen theme highly influences all aspects of the communication system: concept, tools, creativity and media planning.

The Date and Duration of the Event

The date and duration of the event might seem of little importance at first and yet it plays a key role in the success of the event. It is very important to choose a day, or a

C. Guardascione (✉)
Corporate Identity and Communication, Fondazione Idis-Città della Scienza, Naples, Italy
e-mail: guardascione@cittadellascienza.it

A.-M. Bruyas and M. Riccio (eds.), *Science Centres and Science Events*,
DOI 10.1007/978-88-470-2556-1_13, © Springer-Verlag Italia 2013

Fig. 1 Announcement Poster for the Owerri Science Festival

short period, that corresponds with religious or national holidays in order to ensure a potentially higher number of guests.

The communication plan should be created alongside the marketing department and the scientific committee, which should have already defined and developed the majority of its contents. This work should be started three months before the opening of the event, after the theme and the guidelines of the exhibitions have been defined (Fig. 1).

Before defining the strategy of the communication plan, the budget must be established; how much money do we have to organize the communication plan?

Target Audience

The "target audience" may vary while the educational value of the event is consistent each year. For this reason schools usually receive priority. Identifying the "target audience" is fundamental in choosing the "media strategy" and in creating and formulating the message to be transmitted. The educational realm remains our main target. Yet other target audiences may be included over the years with elements used to satisfy cultural curiosity or that allow people to use their free time in a more "intelligent" way. Defining the target audience and the date and

Fig. 2 Futuro Remoto 2011 advertising poster

duration of the event go hand in hand. Since the date and duration are decided beforehand, the target audience must be chosen on the basis of when the event will be held. Date and duration heavily influence collateral targets, and sometimes also the main target audience.

Communication Messages

It is not enough to send one kind of message. The message itself must be designed according to the different targets that have been identified, meaning that more emphasis must be placed on the educational aspects of the event rather than on entertainment features (Fig. 2).

Media Planning

Once the message and the target have been defined, the media to use in accordance with the allocated budget must be chosen.

Billboards, dynamic advertising, radio and the press have proven to be more in tune with communication goals over the years than other kinds of media. Sometimes the local nature of the event may limit communication to a local, provincial and, for some media, regional level.

The message and kind of media must also work in parallel because if the chosen medium influences the message given, then the message can drive the medium we choose.

Communication Tools

Conventional advertising:

- Billboards;
- Posters;
- Flyers;
- Brochures;
- Radio spots;
- TV spots;
- Cinema spots.

Billboards, posters, flyers and brochures should be distributed in crowded places. Radio, TV and cinema spots should be broadcasted during the most watched or listened to programs (Figs. 3, 4, and 5).

- Non-conventional advertising:
- Press office;
- Direct response;
- Participation in trade shows;
- Public Relations (PR);
- Opening ceremony;
- Press conference and other events;
- Web advertising;
- Dynamic advertising.

Dynamic communication techniques will be explained in the chapter entitled "Communication Techniques". However, it should be noted that nowadays the press office plays a key role in determining the success of an event.

Techniques

As previously mentioned, the work of the press office has a primary role in determining the success of the event for two reasons.

The first one consists in having estimated that the potential audience is cultured audience that reads at least one newspaper a day, and follows the television programmes to cultural orientation in addition to news broadcasts. The second reason is more strictly economic as the work of the press office reaches a wide public with a limited budget.

Fig. 3 Flyer for a temporary exhibition on insects

Fig. 4 Programme leaflet for the annual school convention

Fig. 5 Owerri Science Festival programme

The activities for the launch of the event start 3 months before the opening, with a screening of the national press. Then the work consists in contacting individual editors in order to carry out services, articles, reportages.

The press office works towards national and local newspapers, and also to the national and local television, public and private, and the web too.

The Press Release

The 5W: Who did What, Where did he/she do it, When and Why.

Everything in the first lines. Putting the news first is strictly based on a technique that want a bare and *essential prose*, if possible without adjectives; periods of pure information, without interpretations, positions, judgements on the merits; slim and essential texts, which soon come to the point and that open in a captivating way. Everything must be summed up in 15–20 lines.

The *typical syntax* of a journalistic agency is: subject, verb, object, time and place complements and then the other complements. A golden rule for the press officer is contained in the well-known abbreviation A B C, which is equivalent to:

- *Accuracy*: be sure of what he writes, check the information from official sources in order to avoid errors.
- *Brevity and clarity*: use the same language of the reader, which is the journalist, who in turn will speak that of its readers: helping him means improve our result.

The journalist habitually uses a technique of fast reading: he reads as a kind of mental scanner. We must use a graphic form and face out the news: all titles should be written in caps securities, the summaries in italics, the keywords of the text in bold types.

There are reports that may be of interest for their many aspects, in various sections of a newspaper. Some numbers, statistics, percentages, some trend remarks to costume journalists, a particular stimulus of curiosity to who writes about current affairs, something more cultural to who is interested in culture. There is no time enough to write so many different releases on the same news? Then we give everyone the opportunity to deepen the theme that they need. We build the text, concise, complete, and then we attach deepening notes, divided by theme, that the subeditor will select according to its readers. We can send them only with an e-mail or a phone call that anticipates the sending of the release, thus encouraging the welcome and preparing interest.

Online Press Communication

The stages of online press communication plan are simple. In 3 months there will be at least six press releases. We should consider from the beginning that *not all information should be widespread together* in order to have more news to issue a little at a time. Here it is an example.

- The first press release informs about the new event that is coming, so as to create an expectation;
- The second release is the official programme, where there are all the logistics information and how to participate;
- The third and fourth press releases indicate the presence of the sponsor and important testimonials;
- The fifth press release indicates that in the event there will be an important moment that has not been presented yet (e.g. a round table, a game, refreshments, an interview).

Direct Response

It is another technique to make communication happen. Once the target is identified it can be reached by the techniques of "direct response" whose channels are preferential and direct, that are: sending monthly electronic newsletters, sending a

Fig. 6 Stand of Città della Scienza for trade shows

Fig. 7 Opening ceremony in the Newton Hall of Città della Scienza

paper catalogue of activities; finally, a massive telemarketing campaign urges further the target to book tours of the event.

Participation in Trade Shows

The communication can also take place at fairs of similar areas. Participation in events that have theme or target much in line with the event allows an effective and "low cost" spreading of the message (Fig. 6).

P.R. Public Relations

Since the early moments of the conception of each edition a deep "system of relations" with the world of stakeholders related to the subject of the event starts. The entrepreneurial companies (local, national or multinational) also represent ideals partners: they are informed with great advance, invited to collaborate actively in the event, solicited to sponsor. It be careful that the sponsorship may not necessarily be economic but more often "product/service."

Another environment with which it is important to establish or sustain further relations is the world of school, that is the main target of event like this. People are continuously informed on the progress of work feeding in them the "sense of belonging to a community" and often accepting recommendations and suggestions.

Attention: the organizers should have a database with all the teachers contacts (e-mail, addresses...).

Opening Ceremony

The opening ceremony of the event needs particularly care, in the belief that much of the success of the event depends on it. One urges, by sending personal contacts and invitations, the participation of a panel of institutional and/or scientific speakers, a significant political parterre, an audience made up largely of the so-called "civil society" and, last but not least, representatives from school and church (Fig. 7).

Lesson 8: Training Explainers

Fiorella Iuliano

Introduction

Scientific knowledge is often obscure and the same "science producers" are not always able to make it understandable to all. Thus a mediator is needed (journalist, teacher, explainer) in order to convey scientific news in a correct and simple way, taking it from the research world into society. The explainer acts as one of these mediators between science and society.

An explainer works in science museums, science centres, festivals, science direct and outreach activities of research institutions:

- In guided tours of exhibitions;
- As pilots, engaging visitors with hands on exhibits;
- In workshops, laboratories, experimental settings, science shows, pocket science, scientific drama, etc.

The explainer does more than just spread scientific information and guide visitors through an exhibition. An explainer stimulates observation, surprises, arouses curiosity, helps visitors put their experiences in a proper perspective in order to adequately describe phenomena and awakens emotions for better comprehension and to produce a lasting memory. Visitors remain impressed by what an explainer says, does and asks for a longer period of time than they do by the words of an exhibit panel!

F. Iuliano (✉)
Exhibitions and Programs, Science Centre, Fondazione Idis-Città della Scienza, Naples, Italy
e-mail: iuliano@cittadellascienza.it

A.-M. Bruyas and M. Riccio (eds.), *Science Centres and Science Events*,
DOI 10.1007/978-88-470-2556-1_14, © Springer-Verlag Italia 2013

Explainers at Città della Scienza

At Città della Scienza the entertainment coordination staff is composed of a little group of explainers who work both on a permanent or an occasional basis. These explainers satisfy the constant needs of the science centre by welcoming, performing activities for "Sunday visitors", managing exhibition areas, etc. The centres entertainment coordination staff provides the training, evaluation and testing of about 40 people that work on an occasional basis and are managed by an external agency. These people answer to the variable needs of the museum and, in particular, work with school groups. The flow of school groups changes with the season. Some periods in March, April and May produce constant flows with 40 groups per time slot. Entrance to the science centre is divided into five time slots (9.15, 10.30, 12.00 a.m. to 2.00, 3.30 p. m.). Groups of visitors (scholastic or non) can book guided tours or laboratories in one or more time slots. We can handle up to 2,000 students per day.

Most explainers who work on an occasional basis are university students. In fact, we have a continuous turnover of explainers, except for a small group that has been working for us for several years now.

What We Expect from a Science Centre Explainer?

We set high expectations for our explainers, often asking them to cover a variety of positions even in the same day, for example:

- Host – puts on a welcoming smile and explains what there is to see;
- Scientific explainer – explains the science behind the exhibits;
- Creative man – creates new science shows, activities, etc.;
- Supervisor – monitors exhibitions, laboratories and didactic tools;
- Maintenance – repairs exhibits that do not work and discusses upkeep with the relevant departments;
- Technician – starts PCs, software and exhibits;
- Trainer – trains other explainers;
- Security guard – safeguards areas from acts of vandalism, intruders and protects children;
- Writer – communicates and promotes museum activities;
- Source of feedback from visitors – evaluates exhibits and programmes (Fig. 1).

Requirements of explainers are based on the science centre's needs:

- Good scientific knowledge;
- Communication skills – ability to use words, gestures and facial expressions to capture attention;
- Analytic skills – ability to pick up on personality traits and answer the individual needs of a visitor;
- Flexibility – ability to adapt activities to a specific context (target and cultural level of visitors, location, objectives, etc.);

Fig. 1 Explainer with a school group

- Pedagogical skills – ability to make complex notions understandable;
- Management skills – ability to prepare and perform activities;
- Creativity – ability to create new activities;
- Problem-solving skills – ability to handle problems and unforeseen events.

Each Explainer Is Unique

Some explainers may have the same scientific expertise, but may have very different styles in approaching the public, enjoying themselves as they work or in being spontaneous. Their personal skills in communicating science may also vary. No one is perfect at everything and this must be considered when speaking about how explainers are trained and managed.

Is it better to have a team of explainers who each have expertise in a specific field or to have a flexible team of explainers with a wide range of skills? We usually opt for a flexible team of members who have a fair amount of experience in many fields, in order to make the most of the group as a whole (i.e. all of them are able to show the planetarium, but only two explainers are able to perform the show live; all of them are able to guide visitors, but some explainers are better with children, others with disabled visitors, others are better lecturers, etc).

Training Explainers at Città della Scienza

Various techniques are used to train our explainers:

- Senior explainers as instructors;
- Internal training courses;
- External training courses.

In Europe, training by means of senior explainers prevails. Yet at Città della Scienza, all of these techniques are used interchangeably. Explainers at Città della Scienza are trained by an education staff. During periods of lower visitor rates, we collect resumes and make a pre-selection considering the applicant's skills. Young students who have had some experience in coordinating entertainment activities are preferred. Training begins with a short period of apprenticeship. The trainee must learn to face the audience from the very beginning in order to avoid unrealistic expectations. During this short period, a senior explainer flanks the trainee in real work situations so they can experience first-hand the skills the role requires. Training is also supported by written materials and includes courses related to:

- Formal and informal learning;
- Communication science;
- Communication tools;
- Facts and figures about the organization, its goals, activities and procedures.

All trainees have to undergo this part of the training, after which, they can express their preferences. Some prefer working with children or adults while others prefer to work with topics closer to their field of studies.

Training then continues following different paths as the trainees may decide to which courses to take in accordance with their training needs:

- Science Centre – the exhibitions;
- Children's Workshop;
- Didactic activities.

The training ends with an examination. Each explainer must show how he would introduce an exhibit to a group of visitors and is then evaluated in terms of accuracy with regards to the topic, use of communication tools and group management.

Explainers are obliged to participate in new training courses every time a new exhibition is opened. These courses are only to provide information on topics covered in the exhibition as they should have already acquired the appropriate tools.

Some explainers may have extensive knowledge on a number of exhibits, resulting from previous experience or in-house training, but this is not enough. The explainer's ability to involve visitors in using the exhibit so that they may explore the different perspectives an exhibit offers is equally important. The explainer should be able to understand when visitors need special assistance in using exhibits and when they prefer to be left to explore on their own.

Conclusions

What do we expect from an explainer?

What are the needs of an explainer?

The answers to these questions can be used to formulate a training plan that provides:

- Information – facts and figures about the organization, its goals, activities and procedures;
- Rules of conduct – i.e. appropriate attire, behavioural norms when working with visitors;
- Scientific facts about exhibitions and activities;
- Tools – for the purposes of disseminating science to the public.

Lesson 9: Visitor Needs

Francesco De Matteis

Introduction

The visitor is, first of all, a human being. It sounds obvious but we can not forget this little fact especially when referring to the different types of visitors. Visitor needs often merge with the rights of the people. This should be considered in the relationship between the science centre staff and visitors. Human beings are so different from one other that we can not overlook those personal needs that must be recognized and satisfied. We must not forget that a visitor has the right to be satisfied and the science centre staff must be able to recognize the most important needs of each visitor.

Visitors and Diversity

It is helpful to visualize "an image" of visitors in order to characterize the kind of approach that should be taken with them. A science centre is visited by all kinds of people: people of different ages, with different levels of education and different geographical and cultural roots. Visitors also approach a science centre in different ways: in groups, by themselves or with their family. There are school groups and other kinds of groups. The family could be made up of a variety of components in itself. All these characteristics of visitors are combined to create a variable and complex picture that shows us how the needs and desires of those who visit the science centre are different, but also specific.

F. De Matteis (✉)
Exhibitions and Programs, Fondazione Idis-Città della Scienza, Naples, Italy
e-mail: dematteis@cittadellascienza.it

A.-M. Bruyas and M. Riccio (eds.), *Science Centres and Science Events*,
DOI 10.1007/978-88-470-2556-1_15,

There are three perspectives for understanding the needs of a visitor:

Personal Perspective: what the visitor brings to the museum in terms of a psychological make-up of past knowledge, experience, attitudes, motivation, and interests.

Physical Perspective: what the visitor meets in the museum, including the exhibits, architecture and "mood" of the museum.

Social Perspective: interactions between the visitor and friends, other visitors, or museum staff.

Visitor Needs and Motivation

In order to comprehend visitor needs, we have to ask ourselves "why is this person visiting the science centre?"

A museum visit has few practical and directly attributable outcomes. What is the underlying reason that makes someone come to the Science Centre?

More objectively, visitors can be classified into three key clusters depending on the motivation of their visit:

- Intellectually–motivated visitors;
- Emotionally–motivated visitors;
- Socially–motivated visitors.

Different motivations create different needs, but we have to consider all the different motivations combined among them. We can talk about *edutainment*, a term that combines "education" and "entertainment" and it is related to the necessity to satisfy at the same time the cognitive and recreational needs of the visitor. As a consequence, a successful science centre needs to unite the following experiences:

- Learning;
- Aesthetic and emotional delight;
- Recreation and sociability.

Recent studies in Britain show that social motivation ranks number one in a person's decision to visit a museum. It is equally important to imagine the situation in our own country as well.

Welcoming and Informing Visitors

The way in which visitors are greeted upon entering the science centre is particularly important. A visitor who is not made to feel comfortable right from the beginning will probably not enjoy and appreciate the rest of the visit. A visitor must also receive all the information needed to get around comfortably both upon arrival and during the visit. Visitors do not like to get lost. It is also important to let the visitor acknowledge the space around him and get familiar with it, which requires both time and information. In fact, a visitor should never be overloaded with too much information upon arrival, which could make him feel just as lost. In theory, all visitors have the right to have access to the science centre's institutional information. In reality, not too many visitors ask questions about the status of the science centre, staff, etc., but this information must be available to them all the same.

It is important for the visitor to know the centre's 'mission' in order to feel more involved during the visit. For example, when visitors are greeted, they must be made to understand that the staff is ready to work *with* them and not simply *for* them. Likewise, a visitor should have the impression that the science centre is an easy place to get around and any help needed during the visit will readily be available. By communicating this message to the visitors, it will be easier to establish a profitable relationship with them.

As already mentioned, staff must listen to and understand the specific needs of the visitors, which are not always the same. Some visitors simply require information on opening hours, prices, facilities, notable exhibits and navigation aids available (general information).

In addition to this general information, others require more detailed information about the science centre's project and any other information that will help them to plan future visits and to participate in special events (educational information). Still others require detailed information concerning the science centre's exhibitions, collections and links to others sources of information and so on (specialised information). General and educational information should always be given at the same moment it is requested, while it is rather difficult for staff working in the exhibition area to give specialised information. For this reason it is important to be able to contact the science centre's experts when needed.

How Can Information Be Provided?

Basic information can be provided through the use of panels, pamphlets and indications on basic services. However, the physical presence of one or more staff members is needed at the entrance to conciliate visitor needs with what the centre has to offer. An information point is needed. A desk where the visitors can find pamphlets, a programme of events, etc. and where they can meet staff members at the entrance of the science centre is another possibility. This presence is especially fundamental when school groups arrive. In many cases, these groups arrive without a clear idea about what they have to do or can do inside the science centre. It is particularly important that the staff identifies a "reference person" for each group who has been informed of the group's needs.

Visiting the Exhibition Area

Visitor needs may emerge during their visit to the exhibition area as well. In the Italian cultural context, in which people are not used to interacting (especially in a physical way) with exhibits and are used to following a pre-determined path, all kind of visitors (whether they have arrived in search of general, educational or specialised information) generally show a passive attitude with regards to the exhibition area. This typical Italian trend represent a problem when, according to the mission and spirit of science centres of all the world, trying to stimulate the desire to explore and the ability of individual visitors to create their own space of action and own paths. This is true both for individual visitors as well as for school groups. Individual visitors are particularly shy when it comes to interacting with exhibits and teachers require a guided tour.

How Can Visitor's Needs Be Met?

At Città della Scienza, scheduled events may be reserved in advance. For example, schools receive information about the different events being offered either through an event book, internet, a phone call or a visit and the teacher can call our contact centre directly to book the chosen event. Individual visitors and families are generally invited to take part in an exhibit through the use of panels and with the staff's help, but it is also important for them to have the possibility to participate in special events (meeting points, science shows, scientific activities, etc.) that the staff prepares and plans especially during the weekends when these types of visitors are larger in number.

To summarize, the role of the cultural mediator is of a great importance. At Città della Scienza there is a large number of scientific guides to take around school groups and a great number of professional explainers too, specialized in scientific communication, placed in the exhibition areas to work with the public, helping them when it is necessary. The mission and spirit of the science centre – discovering in total freedom – must not be forgotten while visiting the exhibition area. For this reason it is important to underline the role of the explainer, who is not a teacher, but a simple "assistant" and "escort" for the visitors during their exploration. For example, the explainer should not force the visitor to follow a predetermined course nor should he forget that the information provided by the museum depends on the very unique, personal and internal experience of each visitor.

A visitor needs and has the right to feel that his own emotions and thoughts are important. In order to satisfy this need, the explainer must encourage the visitor to be more active and creative while interacting with the exhibit. This is why the explainer should work on emphasizing the visitor's emotions and experiences. In this way, the most satisfying experiences for visitors are those that resonate with their daily experience and provide new information in ways that confirm and enrich their view of the world. The reference to an "every day" phenomenon is more effective than a general and abstract concept.

The explainer must also make sure that people with lower educational levels do not feel discriminated against or lack the confidence to express themselves. Explainers must show a positive attitude that encourages an exchange of ideas between the visitors themselves and between visitors and staff. Explainers must draw attention to instruments that stimulate the public's emotions, thoughts and suggestions. The science centre staff has to be available to listen to the public, apologize in the case of poor service and explain, if possible, the reasons. Explainers have to show interest in suggestions and must never underestimate the visitor's opinions, even when they appear superficial. It is helpful to provide visitors with "comment cards", letting them know that their opinions are important. This also gives them an active role inside the science centre. The 'exhibit' itself is another important issue. Many visitors almost feel lost or fear when they approach an "interactive station". To limit these feelings, explicative panels can be positioned in each station, where visitors can learn to use the exhibit and receive information about the experience the exhibit offers.

The Science Centre: A Comfortable Place

A comfortable space helps the visitor to feel good and learn more. As previously mentioned, the visitor must feel that some important aspects are respected:

- *Cultural*: complete respect for the visitor's educational level;
- *Intellectual*: the use of clear, comprehensive informative and didactic panels, freedom to choose and create their own path;
- *Human*: a helpful and friendly staff available to satisfy general or specific needs, ready to listen and offer support is fundamental;
- *Psychological*: respect for a visitor to take his time, concentrate and think in front of the exhibit;
- *Ethic*: particular attention to not offend different cultures, religions, gender, political beliefs or other sensitive issues;
- *Physical*: a comfortable place that satisfies these basic requirements:
  - Cleanliness not only in the exhibition area, but also for the exhibits and for the instruments used for scientific demonstrations;
  - Proper lighting;
  - Proper temperature and quality of air;
  - Proper acoustics, ensuring that visitors may converse and concentrate.
  - Absence of architectural barriers for the disabled.

Providing basic needs such as food and bathrooms are fundamental in guaranteeing comfort. The exhibits, instead, must be located in a way that ensures easy access. Exhibits should be kept in good running order. If, however, an exhibit is temporarily out of order, the visitor must be informed by a clear and visible announcement.

Changing Needs and Visitor Studies

Society is rapidly changing, creating different visitors with different needs in our science centres. Age has become an important factor in European countries, given our rapidly aging society. Families also present a challenging target group nowadays with the different needs and desires produced by single parent families to patchwork families. The reasons behind a visit to a museum must clearly be seen in a larger social context. Visitor needs have to be considered and studied with dynamic analytical instruments that are fundamental in evaluating the actual change. Two different instruments can be identified for this purpose: one that analyses people who usually visit museums and another that analyses potential visitors that do not normally visit museums, but could be interested in doing so.

Lesson 10: Communication Techniques with Visitors

Davide Petrone

Introduction

For years, educational tour explainers have been present in museums all over the world. Their role is reminiscent of both a scientist and an actor. When the explainer is in fact a scientist too, his job as an educator must not be underestimated only in reason of its being addressed to young school children. It is important that the explainers show their professionalism during both the presentation and the activities. The issues and methodologies regarding their role are briefly described below.

Welcome and Introduction of Schools and Family Groups

First impressions are fundamental as people can typically decide in a few seconds whether or not they like a person or a place. In addition, the explainers must always be aware of their role and maintain proper composure, even during break time, since they are visible to guests at all times. It would be inappropriate being surprised by a visitor while smoking or almost falling asleep while sitting in the museum. The uniform or other distinctive signs allowing an easy identification of museum staff are also very important. For instance, in our museum, for a long time museum employees were only recognizable through a small badge and the visitors mistakenly asked employees of a cooperative (wearing an easily identifiable t-shirt uniform) for information.

A person entering a new place is happier and feels more comfortable if he is cordially welcomed and greeted with smiles. The welcome should not just be a simple 'good morning' or 'good evening,' but should include a moment to introduce

D. Petrone (✉)
Exhibitions and Programs, Science centre, Fondazione Idis-Città della Scienza, Naples, Italy
e-mail: davidepetrone@yahoo.it

A.-M. Bruyas and M. Riccio (eds.), *Science Centres and Science Events*,
DOI 10.1007/978-88-470-2556-1_16, © Springer-Verlag Italia 2013

the museum and to briefly provide information that will help the guest in beginning and developing a proper visit.

The first task an explainer must complete after welcoming a group into the museum is to choose a comfortable space to introduce themselves to the visitor. This is especially important when the museum is crowded. In order to allow everyone to see and hear well, the group should be arranged in a semicircle and the shorter guests should sit in front so their visibility is not impeded by taller visitors. It is always preferable for the explainer to keep the most lively and potentially disturbing children close to him. Additionally, when receiving school groups, it is advisable that the teacher be kept in consideration and be treated with the utmost respect and cordiality.

While moving, if the explainer is familiar enough with the museum, it is a good habit for him to walk backwards without turning his back to the group. This allows him to always keep the group in sight. Also, this increases the children's fondness of the explainer and keeps them more attentive. Children sometimes ask: 'How do you do it?' 'Do you have eyes on the back of your head?'

In addition to introducing themselves and the museum, it is important explaining what the visit will entail and specifying that the activities neither constitute a test or an ordinary lesson (formal and informal communication). In the specific case of our museum, as in other science centres, it is important to specify that it is not an ordinary museum, where touching things is forbidden. Rather it is an open learning environment where it is necessary to touch, to speak, and to participate. Guests should play the role of assistant scientists and act as 'guinea pigs' for the explainer as they try to involve all the children in various activities. If time allows, the explainer should involve all the children by assigning each one a task (particular attention should be paid to students who are either extremely shy or hyperactive).

This is the first act of a performance and it is extremely important to get it right. If the introduction is not well conducted, or the guests cannot hear or see the explainer, or people are inadvertently excluded from the activities, the explainer will pay the consequences for approximately the next hour. These presentation and management procedures are useful for both family groups and school groups, for guided visits, and educational and popular activities of every type.

Scientific Communication Methods

The Voice

The voice is an important tool. During the activities, microphones are often not available and yet an explainer must still make sure to be heard without shouting. Screaming can leave him breathless and makes listening and learning less pleasant. Small and simple exercises, such as those borrowed from the theatre, will improve the ability to positively and effectively communicate in these situations. Learning

Fig. 1 Science demo with the 'Hair-raiser' exhibit

to use the tones, pauses, gestures, and expressions of a true actor will make "lessons" a real performance given to a large audience. It is truly satisfying when the guests applaud to the explainer at the end of an activity and say, 'we thought we were going to have a lesson, instead we assisted to a performance.'

It is natural that, over time, the explanation of certain exhibits, theories, or complex scientific concepts will begin to follow a non-written script repeating the same sentences, jokes, and questions each time. Material, however, must be kept fresh so it does not sound as if one is reciting a poem learned by heart.

The explainer must learn to repeat meaningful sentences every time as if it were the first, using appropriate and humorous tones and pauses. It is a good habit to observe other people working as explainers, leading educational activities, conferences, shows, telecasts, etc., since observed methodologies, expressions, jokes, and movements can then be incorporated into his "own performance" (Fig. 1).

The Body

Other important communication techniques include gestures and poses. Few things are more unpleasant than a person speaking with his back to the audience, or speaking while sitting in a chair and appearing exhausted, or gesturing using frantic movements. It is important to learn to control movements. A visitor will perceive tension if an explainer paces in an uncontrolled manner, rocks back and forth on his feet, or handles objects in a neurotic way. As a consequence, the guide will appear to be insecure and less reliable. The clothing must be appropriate and can often be improved by wearing something distinctive such as a hat or a pin or by hanging a magnifying lens around the neck or a geologist's hammer from the belt.

Last but not least, eye contact is imperative. An explainer must always look at everyone straight in the eye. If he does not keep appropriate eye contact and speaks while looking up in the air or staring at the floor, he will lose visitors' attention. If visitors feel observed and keep eye contact with the explainer, they are more likely to pay attention. It is better to avoid focusing on one individual or group of listeners, however, so the others do not feel excluded. The only way to keep one from being distracted or not listening is to keep everybody in sight.

What to Avoid and What to Point Out

The most intelligent children often have great intuitions and can give correct answers without using the appropriate terms. An explainer should use simple terminology that children understand, as it will help the comprehension of complex concepts. Simplifying concepts is acceptable, but never make false or inaccurate statements. Even a child could unmask a false or inaccurate statement. If this were to occur, nothing of what is said afterwards will be believable. It is better to admit a lack of knowledge about a particular topic by explaining that 'a scientist does not know everything', rather than making false statements. Accurately answering as many questions and curiosities as possible will be useful and gratifying.

Explainers should typically only show experiments they are sure will succeed. It is acceptable to try other experiments, but not too many times, as experiments that are known to fail should not be repeatedly attempted. Never introduce an experiment in this way: "I will now try to show you how this object floats, even though, I have actually never succeeded in doing it... you see, it sunk this time, too". In addition, before moving from one area to another, he should explain what the group is going to do with a simple sentence or joke (Fig. 2).

Some theories will be acted out and told as if they were a story. The explainer should be aware that, as an actor, there is a base script that must not be changed, yet he cannot always expect the public to have the same reactions to the story. As such, in order to keep the script interesting, he must take inspiration from the guests by using their reactions and suggestions to further develop the script and add related wisecracks. Explainers should never be jealous of other colleagues' jokes and wisecracks, or overly attached to the "non-written script" they have created. The explainer should always be willing to change the script by borrowing from others and letting others borrow from them. A person may create a great line, but when repeated by somebody else with a different intonation, it may appear even better. Why then not try and repeat it in this new tone to improve the effect of one's own script? It is, in short, an ever-changing script.

Teachers

The explainer's role is very different from the teacher's one. In a short period of time, he has to present topics aimed to evoke, both in children and adults, new ideas that they can reflect upon during and after the activity. The teachers should prepare

Fig. 2 Explainer with visitors

their pupils prior to the visit or the activity at least by discussing the lesson's subject matter. In addition, the teachers can and should revise the topic when they return to their own classrooms.

The interaction among the two figures is fundamental. Many teachers mistakenly do not prepare prior to the visit or activity and it is important for us to attempt to correct this behaviour. This is actually very difficult. Here are a few common statements with which teachers introduce themselves:

> Well, here we are. These are our pupils. We're going to have some coffee. Good luck. Do you want us to bring you something from the coffee shop? No? Well then, we will be back in an hour.
>
> Well, boys and girls, this is the teacher. Please, do not chat, do not move, do not talk nonsense and, remember, you'll be asked about your visit later.
>
> Don't goof off or you will give a bad impression of me.

Teachers often make such statements in order to appear capable to the explainer. It is recommended to explain to students and teachers that questions are meant to involve the students and to assist in transitioning to the answer. Occasionally, a teacher will feel excluded and, therefore, the explainer should find a role for them. For example, the teacher could be asked for assistance in managing the group or for their involvement in the explanation of the exhibits. It can be fun for the teachers as well as the children if the teachers try the experiments and play the same games as the students. It will make them appear more human and friendly. But the explainer should avoid conflict or competition with the teachers as their role is important to the students. Offending or excluding a teacher is never acceptable.

During classroom lessons children often learn by memorizing definitions, theories, and other sentences without actually understanding them. A good practice could be to have the children recite a fact together, and subsequently the explainer could then re-phrase the statement and, with the aid of other available tools, make it more understandable to children. The same management and relationship methods should be used when training teachers. Too often trainers do not put into practice the theories they explain. Also, people listening to a lesson typically take notes. If the explainer notices that this is negatively impacting student participation or it is slowing down the explanation process, they should suggest (if the teacher agrees) that the students just listen without writing.

Managing and Involving the Group

Both family and school groups are comprised of many members with various ages and personalities. Additionally, there are activities that work well with a large group (e.g. a science show or lecture) and there are others that work well with smaller groups with a maximum of 30 people (e.g. guided tours and educational activities). A perfect number of people for each activity does not exist.

The importance of voice intonation and the proper positioning so everyone can hear the voice has been previously considered and discussed. To be sure of this, the explainer should ask for feedback. He should be careful when asking the group "Can you hear me?". The most frequent answer is "Yeeeeeessssss!!!" but it often comes from someone who is standing nearest to the guide, while the guests that are the furthest away might not even have heard the question. A better idea is to ask everyone to raise their hands if they can clearly hear you. If the explainer can not see all hands raised, then he needs to raise his voice.

As previously mentioned, it is important to always maintain eye contact and to be positioned so that everyone in the group can see you. To keep attention high, for example, some students can be in charge of making sure nobody presses buttons before the moment to do so or that other students are not left behind. Others can be made assistants by letting them carry or hold some tools. The children should also be encouraged to answer and participate in the experiments.

During the guided visits, especially with younger children, it can be of great utility to walk in a specific manner from one exhibit to another. For example, walking like a 'scientist', taking leisurely and thoughtful steps or walking like an animal that was just discussed. This will help to maintain the student's attention while moving from one exhibit to another. If there is one particular child that is taking time and attention away from the other students by always answering first or asking an excessive amount of questions, the child must be gently limited while being careful not to offend or hurt his feelings. Children are clever and can easily understand when they are told 'Let's try to let the others speak too,' or 'I know that you know, but I'd like to hear from the others too.' If someone asks a question while moving from one place to another or when the group is positioned in a manner where not everyone is able to hear, it is better to postpone giving the answer until the whole group can participate. For example, once the group has stopped, the explainer should ask the person to repeat the question or say to the group 'somebody has asked me' and then state the answer so everyone can hear.

The introduction is crucial for large events. The explainer should start by greeting the crowd and in turn the crowd will greet him. If the group's response is not lively, the explainer should have them repeat their greeting until he receives a strong response. This method will help ensure that it is clear the performance has started and will also assist in starting the group's discussion. In addition, it is good practice to select members of the group as 'assistants.' If visitors see their child, nephew, or friend 'on stage', so to speak, they will pay more attention.

Concluding the Visit

Often, the explainer forgets or dismisses the importance of the tour's conclusion. If the guests do not have a clear indication that the show is over, they will remain with the feeling that they would have or should have been able to see more. Also, some people wait for the tour or activity to end before asking a question. As such, it is impolite to immediately disappear at the conclusion of the tour without giving the guests time to ask any remaining questions.

A well-managed conclusion should include some advance warning that the tour is nearing the end.

For example, "this is the last experiment that we will see today." This will give the public the impression that they have completed a well-organized activity with a beginning and an end. It is also customary to gather the group together to re-establish order and to announce the end of the tour or activity with a friendly farewell. This is similar to an actor 'bowing' and if the tour is over, people will likely applaud. For example, "the last thing I have to tell you today is good bye or, even better, see you again. If you have more questions or are curious about anything, I will be around for a few more minutes."

Another good practice is to give suggestions to the guest on how to continue their visit after the tour is over. If there are several groups in the same day that are following the same structure of activities, every group should receive the same treatment.

How to Behave While Waiting and During Critical Moments

In large museums, during the periods when there is a high number of visitors, exhibition areas can get crowded and some explainers wait for their groups near the entrance. In these instances, it is helpful for colleagues to assist each other as much as possible. For example, other colleagues can help keep an area available for the other group.

It is absolutely unacceptable, especially for who work with children, to partake in rude behaviours such as throwing rubbish or cigarette butts on the floor. Additionally, it is preferable avoid smoking in front of children. In addition, explainers must not use curse words and other vulgarities. During the tour, it can be appropriate, useful, and interesting to use a particular accent, dialect, or youngster's slang when communicating with the group. Also, during perceived difficult times, it is important to remember that the group leader or the teacher is putting their trust in the explainer and the institution. If the explainer encounters an issue with a guest, regardless of the reason and regardless of fault, he must attempt to calm everyone down. This includes offering or giving the impression of offering some sort of compensation, such as an extended visit, additional attention, or a special itinerary. The explainer should be very cautious, however, to avoid making false promises.

Experience and Knowledge of the Subject

In general, an expert in a certain area, or particularly fond of a specific topic, will be more capable of easily and clearly explaining complex concepts in comparison to who has only studied the subject for the occasion.

This is only a general rule, however, and is not always the case. The main point of this statement is to stress the fact that explainers should never invent answers and should only discuss concepts they know. Additionally, they must be honest when they do not know the answer to a question. Even the greatest scientist does not know everything. It is always smarter and more educational to answer with 'I apologize. I don't know the answer, but I will try to find out,' or 'I don't know, but I will research it on the web,' or 'I will ask a friend who is an expert on the subject,' or 'I will consult a book,' rather than invent an explanation or pretend the question was never heard. If the explainer does not have thorough knowledge of the subject, repeating a few sentences heard from a colleague or read in non-scientific magazines can easily result in passing on serious misinformation. Again, it is important to remember that an important part of growing as an explainer is to observe how colleagues discuss subjects with their audiences. Finally, he will occasionally make a mistake or feel foolish about something said or done. The important thing is to remain at ease and not become distraught.

Lesson 11: Participatory Tools, the EASW Methodology

Luca Simeone

Introduction

Starting from our experience, the role of participation methodologies in the creation of science events is very useful. In the last years Fondazione Idis was involved in different project both at a national and an international level on this topic.

Working with participation methodologies for us is a must. Participation is related to democracy and freedom of choice. A good process of participation anticipates future problems. Participation is the right way to involve people in a process.

For the creation of science event we prefer to use structured methodologies in particular we appreciated the key point of the *European Awareness Scenario Workshop* (EASW).

The EASW is an initiative of the Directorate *Innovation and SME* of the European Commission, taken in the context of the *Innovation Programme* as a pilot scheme to promote a favourable European social environment for innovation. The EASW methodology has actually been experimented with success in dozens of European cities as a tool to discuss problems related to sustainable local development in urban environments. The EASW is a one or two-days meeting that usually involves around 30/35 people belonging to four different social categories:: residents, policy makers, technology experts and private sector representatives. In general terms, the purpose of the workshop is:

- To exchange knowledge, opinions and ideas among residents, technological experts, policy makers and private sector representatives;
- To identify and discuss differences and similarities of problems and solutions as perceived by the different groups of participants;
- To identify and discuss the major barriers to sustainable urban living;

L. Simeone (✉)
International Projects and Relations, Fondazione Idis-Città della Scienza, Naples, Italy
e-mail: lsimeone@cittadellascienza.it

A.-M. Bruyas and M. Riccio (eds.), *Science Centres and Science Events*, DOI 10.1007/978-88-470-2556-1_17, © Springer-Verlag Italia 2013

Fig. 1 EASW workshop

- To generate new ideas and guidelines for future actions, policies and initiatives both on a local, national as well as the EU level;
- To enhance a public debate in the local community on sustainable urban living in the near future and the role of technology (Fig. 1).

The EASW methodology could be a useful instrument for public participation, as it permits to exchange opinions, discuss visions, identify obstacles to sustainable development as well as solutions pointing in the direction of land ecology. The purpose of the EASW is to promote at a local level a series of meetings and debates which endeavour to assess the possible use of science and technology in terms of implementing plans and programmes for sustainable development. In conclusion, the EASW is an instrument for increasing democratic participation in the choices associated with scientific and technological development. It allows its participants to exchange information and discuss the central themes and processes which govern technological development and the impact of technology on society.

What Is a European Awareness Scenario Workshop?

A *European Awareness Scenario Workshop* is a particularly appropriate means of encouraging the involvement and active participation of social organisations in the passage towards sustainable development. Four different social categories (residents, technologists, local administrators and private sector representatives) take part in a working meeting that lasts one or two days in order that the local community can compare ideas to a common future strategy for the territory.

Fig. 2 EASW workshop

In particular, during a *European Awareness Scenario Workshop*, participants meet to exchange opinions, develop plans and suggest ideas about resolving the problems of the city in which they live; accordingly, they question each other about the role that technological development and systems of social organisation (voluntary bodies, public services etc.) can play in rendering the life of local communities more attuned to the needs of future generations.

The process means that all participants can simultaneously act as beneficiaries and promoters of change while everybody should be considered *experts* since:

- They know local conditions, opportunities for change as well as limitations;
- They can promote change by modifying their own models of behaviour (Fig. 2).

The EASW project has experimented this methodology with success in dozens of European cities (Glasgow, Turin, Lyon, Barcelona, Copenhagen, Munich, Bilbao, Lisbon, Venice, just to cite a few examples). Workshops in these cities represent an important step in promoting greater attention to sustainable development and the role that new technologies, or new systems of social organisation, can play in the solutions of environmental problems as well as for putting forward new ideas to improve the quality of life in local contexts. This is a significant indication of the social needs that have to be met by innovation – both technological or organisational – which can help in local proactive decision-making.

EASW Methodology

It was developed in Denmark after the experience of the Danish Board of Technology and was then made applicable in the whole of Europe thanks to the work of an international group of experts: the Danish Board of Technology, Denmark; Centre for Technology and Policy Studies in Holland; Fondazione IDIS in Italy.

Although the EASWs were created to provoke debate on the theme of urban ecology, they subsequently found applications in a wide range of other areas, from transport problems to the regeneration of disused industrial areas, from new information technology to information about help for the elderly, from information about AIDS to the planning of energy use, to the definition of a plan for local sustainable development.

The workshop is organised around two main activities: *vision making* and *idea generation* and four theme reference areas. In a workshop on urban ecology, for example, the topics suggested for discussion are: (1) solid waste management and recycling; (2) water supply and use, waste water; (3) energy supply and use; (4) daily living and housing.

During *vision making* activity, the participants, after a brief plenary introductory session, are invited to work in four role groups according to the social category they belong to (citizens, administrators, etc.). During the work group, participants are asked to project themes for the future in order to imagine, according to the topics chosen, how to resolve the problems of the city in which they live and work. They must do this using as a reference point the scenarios which offer four possible alternative solutions to the problems under discussion (based on different combinations with regard to the use of technologies and the organisation of social life). In order to facilitate the creation of visions, the methodology involves a series of techniques for organising the discussion and reaching the envisaged results. The visions of the future worked out by each group must be presented at a later plenary session during which an attempt will be made to establish the ideas for the future of all the participants. Each proposal should contain precise details of the solutions to be adopted and underline, in each case, the role played by technology and collective organisation. The final proposal that emerges at the end of this working session will be the basis for a following one.

During *idea generation,* participants are asked to work on four themes which are pre-selected for the workshop; in particular, they are asked to offer concrete proposals as how the suggestions put forward for each theme during the discussion of proposals for the future can be achieved. By working in groups, formed this time according to the topic under discussion (refuse, energy, etc.), participants are asked to put forward ideas about how to achieve the agreed proposals for the future that came out of the first session. Even during this second group of activities, the discussion should be guided with the help of a series of techniques aimed at getting each group to formulate *how* concrete ideas could implement the group proposal and to indicate *who* should take responsibility for their implementation according to the assigned topic. Each group is allowed to formulate a limited number of ideas (usually five). These ideas will be presented in a subsequent plenary session for discussion and, if necessary, will be voted for. The ideas which are selected may eventually become part of a plan of local measures worked out by the participants in order to resolve the problems under discussion.

Fig. 3 EASW workshop with Nigerian students

How Adapt an EASW to Built a Science Event?

The EASW methodology is a way of encouraging democratic participation; it allows participants to exchange information, discuss the issues and processes that determine technological development and its impact on the natural and social environment, aiding the ability to identify and plan real solutions to existing problems.

For the Owerri Science Festival we adapted this methodology, in fact starting from the role groups we decided to create following categories:

- A communication group;
- An education group;
- A design group.

The discussion in these groups was useful to achieve the whole project idea from the partners; at the end of the first session the facilitator shared a common vision originated from every group (Fig. 3).

During the second session the partners worked at the idea fulfilment in three groups:

- Communication;
- Education;
- Exhibition planning and logistics.

This activity was important to define real ideas and the first steps towards the Science Festival. Our experience demonstrated that the EASW methodology is very appreciated for this case study in particular to:

- Allowing the exchange of knowledge, opinions and ideas amongst experts;
- Identifying and discussing similarities and differences in the perception of problems and their potential solutions among the various categories involved;
- Developing new ideas and guidelines;
- Stimulating public debate in the local community.

Lesson 12: Hands-on exhibits based on local handicrafts: the experience of Calitri

Pietro Cerreta

A Science Centre in Calitri, a Little Town Far from the Cities

ScienzaViva is a group of teachers, students, craftsmen and technicians. Their activity in Italy is an original work in the Informal Learning and Public Understanding of Physics and other Sciences. It is a non-profit Association based in Calitri, a little town in Campania, at about 150 km from Naples, in the Ofanto Valley (Fig. 1).

The group is deeply committed in connecting experts, simple students and common people starting from the mismatch existing in science, in particular in Physics. The main activities are the direct production of exhibits and the experimental diffusion of scientific culture. With a genuine educative research, almost independent from the academic streams, ScienzaViva engages people, being living and vital, similar to a *grassroots movement*. ScienzaViva spends the same vigour finding out the deep reasons of the old conflict between formal and informal learning attitudes, going back to the origins of scientific facts.

Thanks to the Calitri Municipality, ScienzaViva is based in a two-storied building where there had been a Technical School. It is the same seat of a Science Centre containing a collection of 50 interactive exhibits and a laboratory for creative workshops of exhibit prototypes. The building is modern, but it is settled at the feet of the ancient village hill in a beautiful characteristic landscape. The Science Centre is available for students, teachers and anyone who is simply curious about science.

Along its 11 years of live, local and national public Institutions have recognized ScienzaViva worth of their consideration.

P. Cerreta (✉)
Associazione Scienza Viva, Centro della Scienza, Calitri, Italy
e-mail: pietro.cerreta@tin.it

A.-M. Bruyas and M. Riccio (eds.), *Science Centres and Science Events*, 145
DOI 10.1007/978-88-470-2556-1_18, © Springer-Verlag Italia 2013

Fig. 1 Visit of Nigerian students in Calitri

A Strategic Alliance Between Teachers and Craftsmen

Our educative model is the Exploratorium of San Francisco, but the principal action fields are schools, students and teachers, as well as common people, students' parents, relatives and friends.

The title of our exhibition is "The square wheels", from the name of one of our most successful exhibits (Fig. 2). The public loves it because it surprises everybody, showing square wheels rolling smoothly as circular ones. "The square wheels" is requested everywhere, travelling along Italy like a Tespi's travelling theatre and it is performed in schools, in museums and in universities.

For this exhibition our association makes use of funds granted by various Institutions as well as by the Italian Ministry for Education, University and Research. The cooperative contribution of IIS "Maffucci" School of Calitri is fundamental too.

ScienzaViva proposes the *hands-on* approach to Science, i.e. it invites to learn science through the use of the hands: touching, manipulating (Fig. 3), exploring. In this way the local handicrafts establish a productive collaboration with teachers. The local craftsmen are in fact a precious source of knowledge. We consider them as "living treasures" because they are irreplaceable for the production of our equipment. Moreover, craftsmen give us a pedagogic model where science is involved in problems more than on apodictic statements learned at university. A synthesis comes out from this "cultural" alliance: scientific and technological, formal and informal knowledge, put together give instruments of rare pedagogic efficacy.

Without such a synthesis, we could not solve the well-known theory-practice clash of scientific education. Other Italian groups cannot get similar results because they are not protagonist of the same building process. Some of them, in fact, prefer to buy their exhibits from international producers. Their instruments certainly satisfy the popular tasks for which they have been created, but they do not give the public the

Fig. 2 Explanation of the 'square wheel' exhibit

idea that everybody can make them, with their own hands, too. Some others propose a science made by "cardboard, adhesive tape and thumbtacks", good for safe places like schoolrooms and libraries, but it does not survive the impact of the great public.

In short, our commitment consists in a series of enterprises: the autonomous production of interactive exhibits, as above mentioned; the introduction into schools of a way of learning coming from science centres from all over the world; the identification of the theoretical dimension of the *hands-on* activities we employ in teaching physics; the travelling scientific presentations, around Italy, that involve "empirical knowledge" in addition to scientific rules and laws.

After all, we prefer being socially and intellectually independent from Universities and from the Big Science and more interested in the practice of common people and their difficulties in understanding science.

The Association is led by people whose main activity is teaching in schools thus guaranteeing a constant curricular lecture of science and not only the science centres' empirical lecture. Moreover our publications guarantee a point of view related to schools, which is hard to find out in the panorama of the specialized publications. Observing it with an expert eye makes the skill of ScienzaViva a bright example of how a science centre created by schoolteachers can be, or how a school can become as a little science centre.

A Little Italian Seed for an African Flower

We immediately accepted the invitation of the friends of Città della Scienza of Naples to generously contribute to the cooperation project for a new Science Centre in Owerri, giving to the Nigerian students the opportunity to study our exhibits.

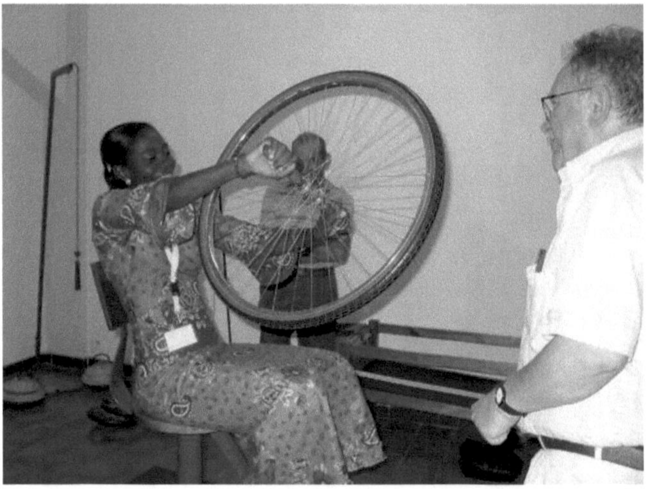

Fig. 3 A Nigerian student experimenting the "gyroscopic bicycle wheel" exhibit

Fig. 4 Scienza Viva Members and Maffucci High School teachers with Nigerian students

As pioneers in the field, we were happy to show our equipment and our laboratory for prototyping exhibits, where we host the teachers which attend our summer schools of interactive science. Moreover, we liked to suggest them the low-cost strategies we have learned to use.

The Nigerian students visited our Science Centre on 9th September 2008. They began their visit at the high school "A.M. Maffucci" of Calitri where the Principal and the teachers warmly welcomed them (Fig. 4).

In the morning, they explored our exhibits, analysing the features and the details; in the afternoon they were involved in the fulfilment of some little exhibits.

Fig. 5 Experiments with optical lenses

One of these has been an optical bench. After having selected a lens among the others of our stock, a group of Nigerian students has learned how to find its focus so that the length of the device could be estimated. Then, they have been guided to observe how to project the layout of the exhibit: i.e. where to put on the exhibit wood basis a little bulb, as a light source, and where a translucent screen, to catch that bulb image. Using two sticks to move respectively the bulb and the screen, they have experienced what is the relationship between the two physical variables of the exhibit. The costs of the material so used were less than fifty Euros, but the evidence of the phenomena was similar to that of a well polished expensive optical bench. We informed the Nigerian student that these activities were inspired to the simple and economic procedures suggested by the recipes of the famous *Snackbook* of the Exploratorium which, being in English, could be easily understood and used by them, too.

The lunch in a typical local restaurant helped all to share feelings of friendship between people coming from parts of world so far from each other, also thanks to the typical spices in the food and the wine.

We hope that the visit of the Nigerian students in Calitri has been a fertile seed for the project of a future Science Centre in Owerri We would like to be remembered by them for the activities observed here. We are ready to supply these girls and boys with all that may be useful to improve their enterprise, if necessary. Our offer has its origin in our roots, the culture of people who have used for a long time their hands with intelligence – i.e. *brain-on*, and we are proud of these feelings (Fig. 5).

Together with me, a physicist and teacher, the members of ScienzaViva who took part to this 1-day meeting were: Canio Lelio Toglia, engineer and teacher, Vincenzo Galgano, secretary of the Association, Vito Antonio Leone, technician and Michele Del Re, artisan.

Lesson 13: Science Theatre

Enrico de Capoa

Introduction

In 1997, a prolific project of contamination between scientific and theatrical languages evolved from a meeting between Città della Scienza of Naples and the theatre group "Le Nuvole", who were called in to manage guided tours and didactic activities.

From Guided Tours to Science Theatre

The first step taken was in imagining that guided tours could become theatrical performances. This is how "theatre-visits", that are valid experimental models of scientific communication, were born. The unique and distinctive characteristic of "theatre-visits" is that they are performed not by actors, but by biology, mathematics, physics and engineering experts who have undergone training at a school of theatre science communication. The scripts performed result from research carried out by scientific experts and professional theatre directors. Special attention is placed on gestures, words, mime, narrative and animation style, with the goal of drawing the audience's attention and stimulating their curiosity (Fig. 1).

Indeed among the major points of contact between the *communicative language of theatre and science*, two points are specifically relevant:

- The theatre – a theatrical event with dialogical interaction, which aims to instill a growing interest around the development of a plot in the public;
- The dissemination of science – curiosity provides a due stimulus for continual introspection and the deepening of one's own knowledge.

E. de Capoa (✉)
Le Nuvole, Naples, Italy
e-mail: info@enricodecapoa.com

A.-M. Bruyas and M. Riccio (eds.), *Science Centres and Science Events*,
DOI 10.1007/978-88-470-2556-1_19, © Springer-Verlag Italia 2013

Fig. 1 Science play 'Perché cielo non cade'

The theatre provides both artistic viewpoints and coup de scene that attract the interest of and involve the audience.

Emotions

With scientific communication, scientific discoveries appeal to emotions, leading to progress in research that may be spread to a vast audience. Concepts belong to scientists, not to actors. Life experiences, recollections, images and emotions belong to everybody. The audience does not learn science during a show or a conducted tour, nor during a moment of animation or a technical demonstration. Images and emotions serve to help the audience remember what has been seen and heard.

So why not exploit life experiences, background, images and emotions?

Debates with the Public

Communicating science in a theatrical and effective way involves acting on learners' emotions in order to transfer concepts. Theatre, through its techniques and languages, acts on our emotional knowledge ensuring that the level of knowledge generated – emotions plus concepts – will be at its maximum level.

The artistic-theatrical language is also particularly useful in provoking debate and confrontation on scientific themes with major social implications. Topics like nanotechnology, nuclear energy, saving energy, reuse and recycling are subjects that have been specifically covered in collaboration between the company's artistic staff and the museum's scientists, which aim at:

Fig. 2 Science play 'Pianeta da favola'

- Exploring possible social consequences;
- Raising debate both between the public and the actors and between the explainers and the didactic experts;
- Raising the audience's awareness about the social implications of new technologies and the importance of information coming from multiple sources;
- Stimulating interest around these topics among those who are less sensitive such as adolescents, the elderly and children at risk;
- Creating an active position and participation from the audience.

Theatre itself has been demonstrated as a valid communication instrument, effective in transmitting scientific and societal themes in that it shows topics from different points of view through:

- Role-play, which enables characters to simultaneously communicate contrasting opinions about the same theme;
- Urging the audience to emphasize the position expressed from one character instead of another;
- Open debate on relevant topics that continues even after the show (Fig. 2).

Theatre as a Training Tool

Theatrical language has also been used over the years to *train museum operators*. Training concentrates on the body as an instrument, on gesture as a language and on space as the venue and aims at:

- Communicating science through the employment of exhibits as symbolic elements around which one can move and tell stories;

- Dramatizing a discovery event or historical-scientific research, from which the visitor – individually or in groups – becomes more participative, thanks to the emotions and associations the communicator is able to generate.

We also need to value the recovery of a storytelling naturalness, which is falling into disuse over the years due to the habit of conventional training to restrict the natural storyteller that is in everyone.

The science of the actor, after all, is the constant variation that he himself carries out on the words, sounds and gestures in order to convey his message in the most incisive way.

In the last few years, the relationship between le Nuvole and Città della Scienza has also brought about the development of *joint initiatives to integrate theatre and shows as key aspects in the design of exhibitions, exhibits or communication projects.*

Working in teams – composed of both artists and museum staff – from the very first phase of the project, allows for better integration and use of available tools. People using different languages are drawn together in the creative process. Their union can result in more than the sum of the single parts. Theatre should interpret the contents proposed by the scientific team in a less literal and more suggestive way. The museum team should also use theatre techniques to design museum and exhibition sets, which, at the same time, provide the sets for the science plays.

Steps Taken to Produce Two Science Shows

"The Cloud of Water and Air" and "9 Months, How Does Life Begin?" for children 4–7 years old:

- The topics that will be "performed" are chosen by the museum staff following feedback from meetings and workshops with teachers and educators.
- Primary school children are targeted at. Teachers mostly request plays on water as a resource and its cycle and a first approach to sexual education, pregnancy and birth.
- The scientific staff specifies a series of fundamental concepts to communicate about these topics.
- The artistic staff proceeds by drawing up the theatrical text on the proposed topic with full, creative freedom, considering the target age, communicative demands, spatial and structural needs and the available budget.
- Narration is considered the most ideal theatrical form for the primary school audience. The mise-en-scène for both performances will involve an actress-communicator and the use of various objects and materials.
- An enclosed structure capable of holding around 15 children at a time is the preferred form of theatrical scenery, particularly indicated both for the magical atmosphere created inside and for its ease of transport and installation. The costumes are then chosen and the soundtrack and light effects are created.

- The show's graphic flyer and the communication program are planned.
- The didactic material to be provided to the classes (games, didactic tools and activities) is prepared.
- A demo of both shows is produced for the museum staff. The communication model is improved on and all new suggestions are discussed in teams.
- The shows are performed for pilot groups chosen among families and school groups coming from different social classes. Evaluators and psychologists also take part in the shows.

Museum theatre is a bridge between science and theatre. It stimulates curiosity and, through a process of dialogue and reflection, leads to understanding.

Lesson 14: Appetizer to the Italian Language and Culture

Franca Sibilio and Rosa Giordano

The Italian language course was one of the complementary activities proposed in the training program for Nigerian students. The course was added to the program to create a moment of exchange, especially from a cultural point of view, with the group of Nigerian students that shares the underlying principle of the entire training experience, training that not only "fills" students with content and pre-packaged indications, but also creates an exchange of information and constant comparison at different levels, thus making the co-organization of the Science Festival in Nigeria both real and feasible.

Organizational needs made it difficult to arrange many meetings (only ten lessons were held). This "time" limitation was overcome by playing on the available "space" and through the use of a theatrical approach. Instead of using a traditional classroom, the informal exhibition area for children within Città della Scienza's Science Centre was chosen as the meeting place. The area includes a ball pool, huge pillows and padded flooring where no shoes are allowed. The message relayed through the choice of this area was reinforced by the fact that no tools such as overhead projectors, books, etc. were used.

Trainers got involved along with the students and lived with them a training experience, divided into ten meetings, with the aim of learning the essential structures of the Italian language (Fig. 1).

The method used was a complete variant to general procedures, especially with regards to short, intensive foreign language courses. Many language courses convey stereotypical forms of expression of basic needs of communication, which are preselected. This method, however, tested by years of experience, puts people into situations primarily involving the body, so they do not need to work in classrooms. This approach necessarily involves the choice not to use any foreign

F. Sibilio (✉) • R. Giordano
CEICC, Municipality of Naples
e-mail: ceicc@comune.napoli.it

A.-M. Bruyas and M. Riccio (eds.), *Science Centres and Science Events*,
DOI 10.1007/978-88-470-2556-1_20, © Springer-Verlag Italia 2013

Fig. 1 Italian language course at CEICC, Naples

language already known (in this case English) as a bridge of communication and provides, instead, the use of body language by the whole class.

The course consisted, then, on a trip in and out of self using all the senses, learning to communicate emotions, feelings, driven by the only motivation to meet the others who, in turn, were approaching us.

The starting point was the common desire to meet and make friends: we introduced ourselves, we played with the letters of the alphabet, we learned to associate pictures with words, to greet and to express our first feelings with simple adjectives: I feel happy, tired, hungry, etc.

The route was developed by discovering language through different senses: we started from taste and smell, experimenting to eat at the same table, to discover old and new scents, to taste some typical plates and to end expressing feelings, memories, associations in simple words. Slowly we got to explore the world of sounds of our surroundings: water in different forms, the sea and its inhabitants, our body, which perceives the external world. Touch was used mainly in contact with the ground: we talked about the country we come from and the host country with the aim to discover the origins of each of us and to look for a closer contact with the Earth. Then from the land we looked up at the sky: constellations, planets, zodiac signs to talk again about themselves but with deeper awareness and already familiar words (Fig. 2).

The path could only end in dreams: we put our dreams together and started to dream of a better future for our Earth and for the human beings. The common thread was, therefore, the verbal and nonverbal communication in order to immerse themselves in the Italian language and to exchange experiences and cultures. The route is well related to the project of a Science Museum in Nigeria which can give benefits to the local community. The participants to the course prepared a public performance using the learned vocabulary and the morphological and syntactic

Fig. 2 Informal setting for Italian language course

structures. They introduced themselves in an original way in order to make people know each one of them in their individuality and share their dreams.

In this involving method, the role and the responsibility of trainers is not to lose sight of the goal, that is learning the language, the Italian language in this case, and eventually to transform this intensive experience into verbal communication that takes into account the rules, the features of a language, no longer a foreign language, because the pupils are no longer foreign speakers of that language, because "being seen by others is to see the truth of others" (Sartre).

Part IV
From Theory to Practice

The Owerri Science Festival

Guglielmo Maglio

After the training of the ten Nigerian students in Naples, we wrongly supposed that great part of the work had been already done; but as we headed off to Owerri, we realised that a deeper training should have been organised in place to ensure the success of the project, not only because of the new explainers involved but because the local settings required an adaptation of the activities and the demonstrations.

Besides, the logistic difficulties and the cultural background suggested to shift the communication strategy towards an approach based more on the "telling tales" than on the "experimental" one. We realised, in fact, that the "human interface", or the explainer role, was much more important in stimulating curiosity compared to what happens in Europe. Visitors (especially older students and adults) tend to be shy in front of the exhibit or the science show, so the role of the explainer is fundamental in breaking this curtain of diffidence or shyness.

Communication channels were also very different from what we previously thought. Typical communication of science events here involves websites, newspapers, and television. Once in Owerri, we realised that the best communication channels were the suggestions given by priests at the end of masses and the music truck, a vehicle carrying a music band on, which drove through the town announcing the opening of festivals. Also the teachers of schools were contacted individually, since we could not count on a centralised authority for spreading the info about the festival in all schools in Owerri.

Local Staff Training

Once Fondazione IDIS staff arrived in Owerri, it was soon clear that a new session of training was necessary. The enthusiasm of the many students who took place at the training sessions was a great evidence of the people's interest in such event. But

G. Maglio (✉)
Exhibitions and Programs, Fondazione Idis-Città della Scienza, Naples, Italy
e-mail: maglio@cittadellascienza.it

A.-M. Bruyas and M. Riccio (eds.), *Science Centres and Science Events*,
DOI 10.1007/978-88-470-2556-1_21, © Springer-Verlag Italia 2013

Fig. 1 Explanations of the chemistry lab, Owerri

nevertheless the training was necessary not only because none of the students had ever performed the work of an explainer, but also because there were many commonplaces and false ideas about some science topics (especially about Ecology), which were dealt in the exhibition and science presentations. Anyway the students proved to be full of enthusiasm and genuine interest for this work, and surely most of them would have been perfect explainers if they had been given the opportunity to work in this field.

Not only the future explainers were trained. A special training session was briefly set up in place for information desk people, for people in charge of multimedia, of equipment maintenance and for human resources managers. Nevertheless due to the constant turnover of people, there are some doubts on the effects of the training.

Anyway all the people trained proved to be very clever and interested in developing this work, so it could be predictable that they would have the opportunity to continue these careers (Figs. 1 and 2).

Exhibition Setup

The Exhibition setup was a crucial point of the project. In order to organize things in the best way and to shorten the work schedule, two technicians from *Eventidue* Company, the builders of the exhibits, were invited to Owerri. Due to the lack of expert assemblers and the lack of tools to use for the settings, this decision proved to be of great efficiency. Unfortunately the container with the exhibits arrived late and the two technicians had to leave earlier, but they left the

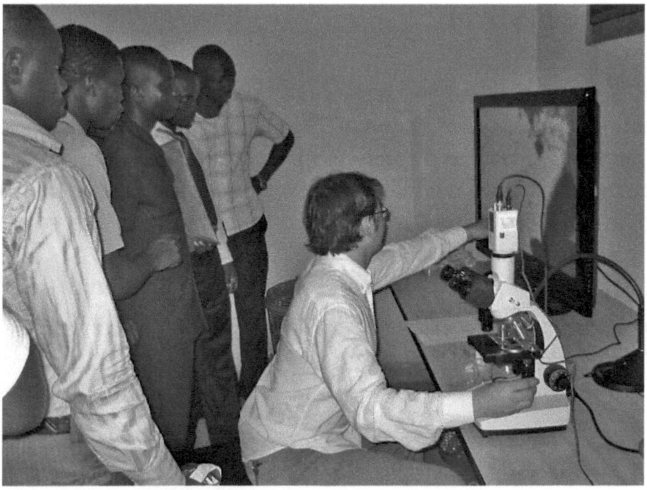

Fig. 2 Observations with the microscope

Fig. 3 Installation of tents for outdoor labs

tools so that the Italian team with the support of the ASCO students could mount the exhibition. Also the problems connected to the lack of electricity could have been solved if the power generator had been sent from Italy, while the local organizers insisted to buy it in Owerri. As a matter of fact, they did not, so we only could organize the exhibition using a small and noisy generator lent by the Archdioceses. The lack of tools and equipment is a situation that must be well considered in similar circumstances (Fig. 3).

Content of the Festival

The Owerri Science Festival has represented the first concrete action of Science Communication towards the Nigerian youngsters and adult people. It was an important step to deepen the awareness of local and international institutions on the importance of science and technology in Nigerian daily life. The festival represented a real answer to the challenge of broadening the scientific culture among the younger generations.

The Festival was organized like a "science fair" with attractions centred on stimulating the curiosity of youths through scientific shows where simple and easily reproducible materials are used.

Hands-on Interactive Exhibition on Physics

By playing with our interactive exhibits, visitors could verify the science facts behind Sound, Vision, Natural Forces and Electricity.

In fact, this interactive exhibition was focused on those science phenomena, which can be detected by the senses and by instruments of observation. It was developed in 4 thematic areas: Optics, Electricity and Magnetism, Forces and Energy, Sound.

All the exhibits allowed visitors to:

- Identify and isolate a set of elements from reality;
- Analyse the relationship between them;
- Translate them into mathematical language;
- Formulate laws which govern phenomena to predict and reproduce them experimentally.

In this way, they gave all the analytical approaches and tools which underpin scientific enquiry. All the phenomena which they could observe and reproduce with the materials, instruments, technical devices and interactive games in this exhibition are those covered by Classical Physics.

Electricity and Magnetism

Electrical and magnetic phenomena have always added a sense of mystery and magic to mankind's relationship with nature:

- Lightning, with its light and the rumble of thunder that follows;
- Magnetite, with its capacity to attract iron fragments and point the North;
- Amber, from whose ancient Greek name, electron.

The study of each of these – and many other – phenomena and their civilian applications of electricity and magnetism underlie some of the most radical changes in the history of mankind. It is a source of progress in every field of social and private life, which gives each of us the power to activate mechanisms, generate light artificially, and extend the time of our activities.

Exhibit displayed:

- Hand-powered battery;
- Electric fleas;
- Imprisoned lightning;
- Solar Radiometer.

Force and Energy

Applying force is a concept that stems naturally from our experience: we apply force when we break a nutshell, lift a brick, model a piece of Plasticine, and wring a floor cloth.

Similarly happens when:

- Wind moves kites and windmills;
- Seawater keeps a boat afloat and smoothes pebbles;
- Magnet attracts a piece of iron;
- Earth makes a compass needle rotate and attracts objects;
- Air supports a hot-air balloon and slows the fall of a parachutist.

We are lighter both in the water and on the Moon, but for different reasons. Which ones? A nail, in water, sinks and a tree-trunk doesn't. Why? Why does a spinning-top not fall over? Why does a stationary bicycle fall over while one that is moving doesn't? Why does the Moon not fall on the Earth? Why is it easier to open a gate by pushing far away from its hinges?

Mankind has established with Nature a relationship made up of continuous attempts to multiply or boost the functions of the human body, thanks to every sort of machine. We have thus succeeded in coexisting with natural events in Earth's life cycles.

Exhibit displayed:

- Archimedean thrust;
- Balancing stick;
- Resonance in a bottle;
- Brick game;
- Chaotic pendulum;
- Squared wheel;
- Gyroscopic wheels.

Light, Colour, Sight

Light has no body, it can't be touched, yet it interacts with matter, which transmits, absorbs, reflects, diffuses, highlights its colours and even, in suitable conditions, generates light. Light makes us able to see, but it cannot be seen. Nothing can move at the speed of light.

Why do house windows look like mirrors in the evening? Why all cats are grey at night? Why a rainbow is curved? Why the sky is black, although the sun is visible, for an astronaut flying in space? Metal can act as a mirror. Also other materials? Why does any object, on reaching incandescence, lose its natural colours and become white and luminous? Why does the surface of a tarmac road seem wet, as if due to a mirage, on a hot day? Why does a branch half immersed in water look broken?

Mankind has always tried to capitalise on this versatile energy, devising:

- Systems and devices to generate and harness it, to have it available even when there are no natural sources (candles, torches, neon lights, light bulbs, laser beams, films, camera bodies);
- Tools to amplify our ability to see (lenses, telescopes, microscopes, spyglasses).

Exhibit displayed:

- Two in one;
- Flying mirrors;
- Newton's disk;
- Mirage.

Music and Sound

There is a close relationship and many affinities between Music and Science. Musicians can be considered as "researchers": they make experiments in order to generate new knowledge. This link is explored here trough a set of experiments, demos, educational games, discussions about three main topics: the physical nature of Sound; the music scales and mathematics; the classification of musical instruments and their construction.

- The physical nature of Sound. The topic is illustrated through experiments and demos focusing on the phenomenon of a sound wave and its propagation, the parameters of height and intensity, resonance and beats.
- The music scales and mathematics. The music scales in the different cultures are all based on a scientific foundation concerning numerical ratios between different frequencies. The numerical basis of the music systems is illustrated to the audience through experiments and figurative elements (lines, points and complex figures).
- The organology. The raisons of the instruments' different timbre production is examined through the construction of simple music instruments: the materials used (wood, leather, metal, etc.), thickness and tension (leather and ropes), dimension.

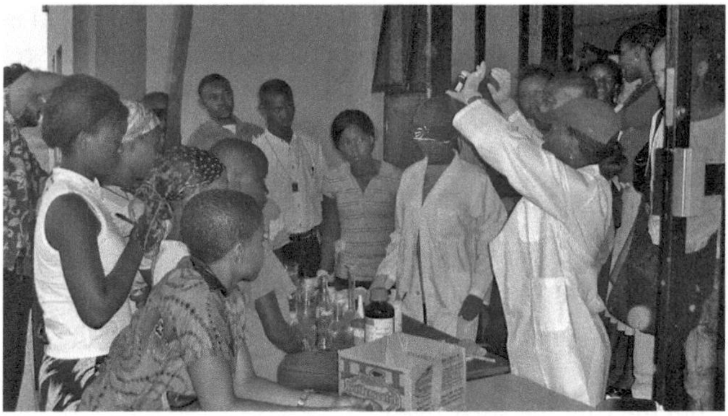

Fig. 4 Outdoor labs

Exhibit displayed:

- Delayed sound;
- Inversion sounds machine;
- Cable telephone;
- Wave visualization.

Activities calendar:

- Sand in resonance;
- Waves running on the wire;
- Count the waves;
- Boing boing;
- Rule and laser;
- Voices and sounds at the computer;
- The waves ring-a-ringo'-roses;
- Big sound and little sound.

Outdoor Laboratories

Around the interactive exhibition there were five open laboratories – Energy, Environment, Music, Chemistry, Food and Health aiming at showing people that science facts behind our everyday actions can be easily understandable and replicable with simple materials (Fig. 4).

Energy

Here we focused on energy and its environmentally sustainable use. Visitors could see a series of experiments and demos aiming to show how energy is produced by Nature; for instance the photosynthesis process, on which many natural processes are based, but also how energy is produced from fossil fuels (gas, oil and coal) and to the physical and chemical transformation taking place during the refining cycle. Here are also introduced the "new" energy sources. A set of experiments will then focus on the relationship between strengths and energy in Nature.

Activities calendar:

- Energy from oil;
- Heat absorption: a black and white issue;
- The wool as a dam;
- Make a thermoscope!
- Make a calorimeter!
- Energy from the sun: the solar oven;
- The thermometer;
- Pet ice cube;
- Is it warmer the wood or the metal?
- Ice cube melting on polystyrene and on metal;
- Experiences on the refraction;
- Greenhouse effect;
- Water cycle in a jar.

Environment

In this area, experiments and information enabled visitors to better understand the natural environment, and stimulate the youth's curiosity towards natural phenomena and everyday issues. Even "hard" topics like the Bio-geochemical cycles (water, nitrogen and carbon) can be easily displayed using simple materials. This helps us to understand how is important to recycle and reuse most of the materials we handle every day.

Here we could also investigate some of the Nigerian natural environments: the coasts, the delta, the bush and the rain forest together with the typical organisms living there. The specific features of the different ecosystems are presented in terms of biodiversity and environmental value. But also the safeguard of local animal species and the fight against parasites and plants and animal diseases is dealt with. The management of resources is also tackled: the deforestation and the subsequent soil erosion, the use of mono-cultivations which spoil the soil, air and water pollution which affects plants and animals.

Activities calendar:

- Fighting plants' enemies;
- Observing cells;
- Life in a water drop;
- Friendly roots;
- The colours of leaves;
- Pollination;

- Water cycle in a jar;
- Soil with raincoat;
- Distil water;
- Please pass the carbon;
- The oil suffocates;
- Oil and bird;
- The worm eats my garbage!
- Ecosystems;
- Baby bean;
- Where are the roots?
- Coloured celery;
- A flower opens;
- Tut the mold;
- Let's prepare a mold house;
- Mimicry.

Food and Health

Here we introduced the matter of a correct diet for a healthy lifestyle. Particular attention is paid to the micro-organisms (bacteria, fungi and parasites) and the viruses which pose a threat to the human health. Animations and simulations show how they affect our body and which precautions need to be taken in order to reduce the impact of these invisible enemies.

Scientific experiments show how the human body works. Simple experiments demonstrate how to determine the content of sugars and starches in food and how to extract (and how to make it visible) the DNA of some food like onions, bananas and tomatoes.

Activities calendar:

- Eat well and stay well;
- How to preserve food?
- A journey through the digestive system;
- Virus and bacteria;
- Build the nutrition pyramid!
- Let's go shopping food so to make breakfast;
- Lunch, dinner and snack.

The Chemistry Bar

A tent hosted the "chemistry bar", where visitors could see easy chemical experiments. Experiments are shown in an entertaining and accessible way, with spectacular effects in order to let visitors understand the processes behind chemical reactions and to formulate and verify hypothesis about the same reactions.

The demonstrations include mainly natural compounds, avoiding complex and polluting substances, in order to facilitate the discharge of wastes and also to emphasize that chemical reactions occur every day in natural world.

Activities calendar:

- Balls dancing;
- Almond's milk – a summer drink!
- Vegetables and nobility;
- The naked egg and the dressed egg;
- At the end . . . a coffee!
- Discovering the red onion;
- Blue for agitation.

Report of the Festival Day by Day prepared with Sylvia Anyadoh, ASCO Owerri

20th April

The Italian delegation arrived in Owerri. The delegation was formed by Vincenzo Lipardi, Fondazione IDIS Executive Committee and President of Ecsite (the European Network of Science Centres and Museums), Guglielmo Maglio in charge of Exhibition Areas and Davide Petrone, explainers from the Science Centre of Città della Scienza, Christian Tasco and Philip Larosa of the Eventidue Company.

21st April

This day was scheduled to be the first day of training for the explainers but since the container with the exhibition and the materials for the science demonstrations had been stopped in Lagos due to bureaucratic procedures, it was impossible to do the training of the students. The exhibition, that wanted to be a positive spur for young generations to approach science and technology, was made of 21 interactive exhibits on different topics (Energy, Music, Light and Electricity) and it was completed by seven computers with multimedia programs and two microscopes equipped with camera that projected pictures on a giant screen, so that the images shown could be observed by a whole school group.

Then, outside the centre, five thematic tents (on Food, Music, Environment, Energy, Chemistry) gave the opportunity to visitors to see science demonstrations using simple and reusable materials. There were a lot of Nigerian students interested in taking part to the training, mostly coming from scientific universities in Owerri, and so we decided to divide them into groups and proceeded with different training sessions on the role of explainers and tested the communication skills of the students.

Guglielmo Maglio and Davide Petrone met a first group of 25 students and talked about the role of explainers, the best way to communicate science and how to manage a group of students in a guided tour. They also illustrated the layout of the exhibition to the students.

They then showed how science demonstration works, using materials and animal/vegetal specimens gathered in the garden of the Pastoral Centre, that were going to be object of naturalistic observation later.

22nd April

The collection of animal/vegetal specimens went on. Furthermore the delegates of Fondazione IDIS-Città della Scienza checked the garden where the tents were going to be installed.

The staff of Fondazione IDIS-Città della Scienza met a second group of 30 students and talked about the role of explainers, the best way to communicate science and how to manage a group of students in a guided tour. They also illustrated the layout of the exhibition to the students. In the afternoon the tents were installed into the football field and the grass was cut.

23rd April

The container with the exhibits and the science demonstration materials had not arrived yet; meanwhile part of the wooden furniture – built locally – arrived. The materials were made of solid wood and seemed to be resistant but with a different size, so the Italian delegates worked to readapt the organization of the tents to use the furniture in the best way.

A meeting between the Italian delegates and ASCO Nigeria was organized in order to establish the different roles and responsibilities of people during the Science Festival.

A truck with a musical band was going to be used by the ASCO association to advertise the opening of the Festival in Owerri. In spite of the delay of the container, the local organizers did not take in consideration the idea to postpone the launch of the Festival, scheduled for the following day (Fig. 5).

24th April

The launch of the Owerri Science Festival, with an official opening ceremony, was first held at the Maria Assumpta Cathedral in Owerri, where the Metropolitan Archbishop Rev. J.V. Obinna introduced the importance of the Festival as

Fig. 5 Festival advertising band, Owerri

a way to promote scientific culture in the local population. Rev. Obinna remembered that during the so called Biafra war, the catholic population was able to resist for many years thanks to the advancement in scientific knowledge, which was very high at the time, and helped people to improve the outcomes of agriculture, food supplies and breeding of domestic animals. This high level of science and technology was possible thanks to the quality of science teachers, both at university and in secondary schools that were the real engine for the development of the region. According to the Archbishop, after the war, this high level of science and technology background collapsed, and slowly the region lost its prominence in scientific education, and this developed in decreasing the life quality and in the use of unsustainable practices in agriculture, industry, energy.

The Archbishop hoped that this Science Festival (and the Science Centre that would come after it) could be a boost to invert this negative process and stimulate young students to access to scientific careers to become the science teachers and professors of tomorrow. Then the opening ceremony moved to the Assumpta Conference Centre Owerri, were the chairman Prof. Chuks Ogbonnaya introduced the participants: Metropolitan Archibshop Rev. J.V. Obinna; Rev. Fr. Dr. Tobechi Anyadike, Director Assumpta Science Center Owerri; Dr. Vincenzo Lipardi, ExEc-utive Committee of Fondazione IDIS-Città della Scienza and President ECSITE; Prof. Anthony Okere, Chairman Scientific Committee Asco Nigeria; Dr. Ngu, Director of Unesco Nigeria, and Dr. Riccardo Rovere, coordinator of cultural and social activities from Finmeccanica, the main sponsor of the event. After the presentation, a concert of traditional music took place in the conference hall. The container had not arrived yet, but the speakers went to inspect the field where the activities would be held.

25th April

Since the container with the exhibition and the materials for the science demonstration was still in Lagos due to bureaucratic problems, the Festival started only with events and conference, without the exhibits.

A local young researcher brought some scale models of caterpillars and truck working with electrical power for a demonstration on how these machines worked. This made it clear that the festival could become a catalyst for activities and jobs related to science and mechanics.

In the first lecture at the Assumpta science festival Owerri Professor N. Onuchukwu presented the improvements chemistry can supply to our daily life. Strategic issues were discussed like producing bio-gas from natural sources, making soap and killing mosquitos in an easy way. After the seminar, Nigerian ASCO students were trained by explainers from Fondazione IDIS-Città della Scienza on communication skills and were asked to explain the audience things they found in the environment. It is clear that this training on the field had a special value, not only for the students, who could interact with trainers, but also for the trainers themselves, who could realize the different meanings of things, the commonplaces, the things to be considered and also the lacks and the positive aspects of the students scientific backgrounds.

26th April

The Italian delegation was invited to attend the traditional mass in a village not far from Owerri and meet the king of the local community. Then they were invited to see the site where the future Assumpta Science Centre was going to be built. At the moment the area was planted with oil palm and cassava, but the king assured that the area was going to be cleaned in the foreseen times. Then the delegation was hosted in the palace of the local community king where the traditional ceremony of the kola nut took place.

27th April

The topic of the first lecture, given by professor Ogbonna, was: "Food and your health". He started by defining nutrition, which according to him simply meant eating what you wanted to stay alive. Size and weight were said to be the characteristics of nutritional adequacy. He likened the body to a machine that receives input as food and output as waste products. He said that the ability of the body to withstand stress also depended on the nutritional adequacy. Nutrients requirement and energy needs of the body, carbohydrate metabolism disease microbes in foods like yeast, bacteria, and moulds were among the issues discussed in the conference. At the end of the lecture, participants were given the opportunity to ask questions which were also answered accordingly by the lecturer.

The second lecture was leaded by Mrs. Professor Onwulili on the topic "science and the preservation of traditional music". Before the lecturer started there was an interlude of a song presentation by FUTO students. In her lecture, Mrs. Professor Onwulili explained how music related to physics saying that music consisted of sound and motion. She listed some traditional media of information transmission like the membranophones and the idiophones. She also discussed the uses of the traditional music for entertainment, to pass on information, to announce death, to sing praise and to communicate. Talking about music and culture preservation, she stressed the need to preserve the Nigerian (and especially Imo) fast-going culture because according to her all the countries in the world had their tradition and so Nigerians should not allow their culture and tradition to fade away. Professor Onwulili talked about the scientific tools to store music, educate the masses on the need to preserve traditional music and the incentives for it, and the use of DVD's CD'S etc.

After the lecture a drama was performed by ASCO students in Imo State Polytechnic Umuagwo. After the lectures the staff from Fondazione IDIS-Città della Scienza gave a special training session on communication skill in a science centre. The students were asked to explain, with simple words and active participation, things around them. The aim of the training was to know how good explainers would be in the transmission of information to the visitor of the science centre and how important they would be in stimulating visitors to science topics.

28th April

The first lecture was given by Professor F.C. Eze on "Alternative energy option for Nigeria and Electricity generation from nuclear power". The features of nuclear power, the solar energy, its application and the generations of solar cells and their features were among the things discussed by the lecturer. Questions were asked by the audience and were also accordingly answered by the lecturer.

Then there has been an interlude of a music presentation by an ASCO student from FUTO which was a real entertainment to the audience.

The second lecture, presented by Prof. Ogwuegbu, concerned the Electricity generation from Nuclear power.

After the seminar, the training of students by the Italian staff of Fondazione IDIS Città della Scienza continued introducing the emotional aspects related to the work of the explainers. In the training session participants were asked to talk about an object which was dear to them, trying to communicate the emotions they were feeling for those objects and to improve communication skills towards other people. The aim of this training, according to Guglielmo Maglio was to show that, talking to other people, explainers could gain a lot in communication skills if they put more passion in the things they were communicating.

29th April

The container finally arrived! At 9.30 the operations of unloading started and the exhibits were put in place. Then the team of Fondazione IDIS – Città della Scienza, helped by the Italian ASCO students, started to mount the exhibit according to the layouts. The mounting operations continued all day long until 11.00 p.m. The hands-on exhibits were mounted first, then all the computers were installed and prepared in a specific room and finally the microscopes with cameras were mounted. Then the huge panels were set.

Professor Anthony Onuchukwu held the seminar of the day on the topic: the need of formulation of chemical products in our homes for economic empowerment.

30th April

The Exhibition was mounted and students liked it very much. They enjoyed interacting with exhibits. Also young students, who were there for the conferences, seemed to enjoy this new kind of approach to science, so different from what they were used to. At first the public seemed to be shy and did not interact much with the exhibits, but after seeing how they worked, they kept using them repeatedly. Especially the "Two in one", the "Imprisoned lightings" and the "Delayed sounds" exhibits amazed the public, but all the exhibition seemed to be appreciated. Also the graphic panels provoked a positive feeling in all the people who saw them.

The Italian delegates trained the explainers, divided in two groups, organized according to the communication skills of the evaluated students. Both the first and the second group were trained to lead a guided tour through the exhibition area, to use the microscopes and to conduct scientific demonstration in chemistry, energy, environment, food and music. All the students were provided with a specific 25-page information handbook on how to make the demonstration. So during the training these instructions were read in public in order to see if everything was clear. The training started in the morning and continued all day. A visit to the institute Don Guanella – that hosted disabled youth – was organized in the evening.

1st May

At the beginning of the day, Guglielmo Maglio helped ASCO Nigeria to organize the basic services for the festival: the info desk, the security staff, the cleaners, the electrician and the role of the explainers. Then the explainers of the first group held science demonstrations in the canopies under the supervision of Davide Petrone while the explainers of the second group led some guided tour of the exhibition under the supervision of Guglielmo Maglio.

Guglielmo Maglio, together with Sylvester, also checked the building where part of the science exhibition would be moved after the festival (especially the computers and the microscope), due to security reasons. The closing of the science festival was postponed to the 4th of May, to make up for the days in which the exhibition could not be mounted.

After lunch a farewell ceremony was set up between the people from ASCO Nigeria and the delegates of Fondazione IDIS-Città della Scienza and some gifts were given to all the people that took part to the project. The Italian delegates left Owerri to return to Italy.

2nd May

Tour guides came in early and were grouped into five according to the thematic areas. They collected and set the materials for the chosen experiments respectively and practiced accordingly. At about 11 a.m. people went upstairs to watch a documentary on African wild life, Lions.

While this was going on, a team of lecturers from the department of Public Health FUTO came around and expressed admiration for the microscopes, Sylvia Anyadoh showed them how the microscopes worked both for three dimensional objects and microscopic organisms using insects and life in a drop of water to their delight. They promised to come back with their students.

Some school groups came in with their teachers and the tour guides took them round the hands-on exhibits in three groups. After the documentary, the experiments were going to continue but rainfall came and people ran back to the conference centre hall (exhibition area) while the school groups ran into their buses and left, some children also dared the rain and went home.

3rd May

An extraordinary meeting of the tour guides was held with Rev. Fr. Tobechi and the secretariat. This was mainly for familiarization and identification of the guides. About 43 people were present. After the meeting the guides continued to familiarize with the exhibits and left afterwards as there was no programme for the day.

4th May

From 8.30 a.m., tour guides in different thematic areas set up their tables and started rehearsing on the experiments to show the visitors.

At 10 a.m. Rev. Fr. Tobechi and two Monsignors officiated the Mass for the closing ceremony. After the Mass everybody went to the conference centre for the formal closing ceremony. The hall was filled with people. The commissioner for Works who was going to chair the ceremony sent her apologies therefore, Prof. Chuks Ogbonnaya chaired while Sylvia Anyadoh moderated as master of ceremony. Many dignitaries were present among which were Msgr. Akagha, Msgr. Aghaizu, Rev. Fr Tony Njoku representing the Archbishop, A.J.V. Obinna, Prof. Iheonye of School of Agriculture FUTO, Mrs. D.I. Mgbeahuru of Imo State, Basic Universal Education Board (SUBEB), Mrs. Njemanze of Owerri Municipal Council, Mr. Saikat Ganguly of NIIT Owerri, Mr. Marco Schmidtt of Switzerland, Mr. Sabino Iandiro & Mr. Glauco of ASCO Italy, many science supervisors from different local government areas in Imo state, some school groups and their teachers, undergraduates, children etc. The press was featured by DAAR communications owner of African Independent Television (AIT), the Leader newspaper of Assumpta Press, Guardian, Citizen and Sun newspapers.

After the closing prayers, everyone went out for the release of the ASCO balloon which flew away never to return to us. This was immediately followed by photographs. The next event was the tour of the hands-on exhibit, one group for VIPs and other groups for younger people. There were five groups in all. Presentation of the temporary site was done by Fr. Tobechi by pointing at it from the corridor of the conference centre (because rainfall did not allow us to walk across) to the admiration of all while thanking the Archbishop and the Archdiocese for the provision.

Outdoor experiments started in the five thematic areas with the tables along the corridors (front corridor; Environment and Chemistry, side corridor; Music and Food & Health), Energy was inside, close to the reception table. There were crowds of people at all the tables enjoying the experiments. Miss Adaeze Attamah supervised that the crowds were contained and orderly too.

The peak of the day was the arrival of the Archbishop, A.J.V. Obinna with Monsignor Amakiri in the evening as the outdoor experiments were on. He first stopped at the Food and Health section where his body Mass Index (BMI) was checked and found to be normal. He was then led around the hands-on exhibits. Finally, he called at the microscope area where he was shown some macro – (different insects) and microorganisms (life in a drop of water). He expressed admiration and appreciation for the Italian counterparts and everybody who contributed to the success of the festival and the science centre. He suggested and prayed that the project/centre would be supported and updated from time to time as well as the tour guides would be trained also on personal research.

Hands-on exhibition and experiments then continued until it was time for departure of two of our friends, Mr. Marco Schmidtt from Switzerland and Mr. Glauco of ASCO Italy. A short farewell ceremony which included exchange of gifts and taking of photographs took place before they left to catch their flight.

The festival finally successfully closed officially at about 6.30 p.m. on Monday, 4th May, 2009. The movement of the exhibits to the temporary site started the next day and now all have been safely moved to the temporary site.

Part V
Technical Annex 1, Cookbook of Science Demos

General advices

Guglielmo Maglio

General Advices

Performing scientific experiments can lead to injuries or damages if done in a wrong way. It is suggested then to require the help and the supervision of an adult (teacher or parent) in performing the following experiments. Fondazione Idis is not responsible for damages that eventually can occur as a result of your failure to follow the safety rules and the instructions included into the description of activities.

Some Safety Rules

- Never perform any of these experiments without the supervision of an adult (teacher or parent)!
- Always wear safety goggles when using chemicals or other substances that might get into your eyes!
- Wash your hands before and after each experiment!
- Never mix chemicals without an adult's permission!
- Be sure that bottles and containers are capped when not in use!
- Immediately notify the teacher or an adult if any chemical gets in contact with skin or clothing!
- Do not look directly into a test tube or smell it when mixing chemicals!
- Never taste a chemical substance you are using or producing in these experiments!
- Clean up the desk and all the equipment after the activity!
- Notify an adult immediately if you get injured when performing an experiment!

G. Maglio (✉)
Exhibitions and Programs, Science Centre, Fondazione Idis-Città della Scienza, Naples, Italy
e-mail: maglio@cittadellascienza.it

A.-M. Bruyas and M. Riccio (eds.), *Science Centres and Science Events*,
DOI 10.1007/978-88-470-2556-1_22, © Springer-Verlag Italia 2013

Workshop 1: Chemistry

Giulia Gaeta, Elisa Rampone, and Federico Brigida

Dancing Balls

Purpose. To show the chemical reaction between an acid compound and a basic compound resulting in the formation of carbon dioxide. Concepts of discussion: density, pressure, hydrostatic force, solubility.

Materials needed. Mothballs, sodium bicarbonate ($NaHCO_3$), water, a dye or an acid–base indicator, wine vinegar or diluted acetic acid (CH_3COOH), a dropper or a spatula or a spoon, blue and red litmus papers, a beaker (5 l), a glass stick.

Preparation and procedure. Pour two spoonfuls of $NaHCO_3$ into the beaker. Add water, a dye or an acid–base indicator and the mothballs into the beaker. In this step the mothballs are at the bottom of the beaker. Add 50 cc of wine vinegar or diluted acetic acid. The mothballs go up and down time after time in the solution. If the mothballs do not rise, add more vinegar.

What's happening? The density of the naphthalene is higher than the density of the water and the mothballs lie at the bottom. The wine vinegar contains acetic acid (test the acidity with the blue litmus paper). The sodium bicarbonate in water produces a basic reaction (test the alkalinity with the red litmus paper). The acetic acid reacts with the sodium bicarbonate and produces carbon dioxide:

$$NaHCO_3 + CH_3COOH \rightarrow CH_3COONa + H_2O + CO_2$$

The little carbon dioxide bubbles drag up the mothballs. When the carbon dioxide bubbles arrive on the surface of the solution, they break and the mothballs start to fall down. During their fall, the mothballs meet other carbon dioxide bubbles and capture them to go up again.

G. Gaeta (✉) • E. Rampone • F. Brigida
Science Teachers cooperating with Fondazione Idis-Città della Scienza, Naples, Italy
e-mail: eliramp@libero.it

A.-M. Bruyas and M. Riccio (eds.), *Science Centres and Science Events*,
DOI 10.1007/978-88-470-2556-1_23, © Springer-Verlag Italia 2013

Almond Milk: But Not to Drink!

Purpose. To show a hydrolysis reaction and to talk about chemical equilibrium.

Materials needed. A jar, a glass, a volumetric pipet of 3 ml, a 0,15M $BiCl_3$ acid solution (by HCl), laboratory gloves and glasses.

Preparation. Before meeting the audience pour a few millilitres of $BiCl_3$ into the glass.

Procedure. Fill the jar with water. Pour out the water from the jar in the glass. The water becomes lactescent like almond milk.

What's happening?

The $BiCl_3$ hydrolyze:

$$Bi^{+3} + H_2O + Cl^- \rightarrow BiOCl\downarrow + 2H^+$$

Or

$$BiCl_3 + H_2O \leftrightarrow BiOCl + 2HCl$$

The precipitation of BiOCl is responsible of the whitish colour.

Vegetables and Nobility

Purpose. To show iodine reactions.

Materials needed. A potato, a tomato, a knife, a small plate, a dropper, tincture of iodine.

Preparation and procedure. Cut a potato into two halves and drop a few drops of tincture of iodine on the surface of the potato. The surface becomes blue because of the reaction between the iodine and the potato amylase. Repeat the experience with the tomato. In this case, we do not get a blue coloration because tomatoes do not contain starch.

Blue for Agitation

Purpose. To show an oxidation-reduction reaction and chemical indicators. To talk about the chemical equilibrium and about the reduction power of sugar.

Materials needed. Erlenmeyer flask (with the cover), a spatula, solid glucose, a 0.1% hydroalcoholic blue methylene solution, a 8M NaOH solution (be careful using it – can cause severe burns!), distilled water.

Preparation and Procedure. Pour 200 ml of distilled water into the flask. Add in this order:

- 20–25 ml of NaOH solution;
- A teaspoonful of glucose;
- 1 ml of blue methylene solution.

Cork the flask and leave it on the table. In a short time the solution changes colour. At first it turns blue. After a few minutes it becomes clear. Shake the solution. It turns blue again. The colour blue is due to the oxidation reaction of the methylene with the O_2 in the air. The colour blue disappears when the solution is at rest because of the reduction reaction of the methylene.

What's happening? The glucose oxidizes to gluconic acid. In alkaline environment the gluconic acid turns into sodium gluconate and the blue methylene reduces itself into a colourless compound. The reduction reaction releases oxygen, which is concentrated at the top of the flask. When the solution is shaken, the oxygen is mixed once again with the solution and the methylene is oxidized once more.

The 'Clothed' Egg and the 'Unclothed' Egg

Purpose. To introduce the concepts of osmotic pressure and density with laboratory experiments.

Materials needed. A fresh egg, an old egg, concentrated HCl solution 2M (be careful using it – can cause severe burns!), a spoon, three beakers, NaCl solution 2M, distilled water, laboratory gloves and glasses.

Phase 1

Preparation and procedure. Put the fresh egg in a beaker. Pour the HCl water solution in the beaker until it covers the egg. Little bubbles of gas will form on the eggshell and then slowly dissolve away. Acetic acid can be used instead of hydrochloric acid, but note that phase one will take longer as acetic acid is a weaker acid.

What's happening? The eggshell is made of calcium carbonate, which reacts with hydrochloric acid:

$$CaCO_3 + 2HCl \rightarrow CaCl_2 + H_2O + CO_2$$

The carbon dioxide forms gas bubbles, whereas the calcium chloride is freed within the solution. Now the egg is 'unclothed' and can be used to talk about osmotic pressure.

Phase 2

Preparation and procedure. Move the egg from the HCl solution to the concentrated NaCl solution. At first the egg will float; after some minutes the surface of the egg becomes wrinkled and the egg drifts down to the bottom of the beaker.

Move the wrinkled egg to a beaker full of distilled water. At first the egg drifts down to the bottom of the beaker, but after some minutes the egg swells again and the surface becomes smooth once more.

Repeat the steps in phase 2 using an older egg that, however, has its shell intact. Important differences can be observed with respect to the fresher egg. In the concentred NaCl solution, the egg with the eggshell will float indefinitely, whereas in the distilled water, that egg will stay on the bottom of the beaker indefinitely.

What's happening? There is a semipermeable membrane under the eggshell which remains impermeable for most solutes, but it does let water pass. A hydrogen bond is found between the water molecules and each water molecule can form up four hydrogen bonds with other water molecules. A solute, especially an ionic one, breaks many hydrogen bonds, but hydrogen bonds tend to form again. The more concentrated the solution, the larger the number of hydrogen bonds that are re-formed. Thus, water tends to go where the concentration of the solution is higher. When the water enters the egg, the air bubble becomes smaller and the opposite happens when the water goes out. The egg floats when the air bubble is bigger and sinks when the air bubble is smaller.

The differences between the behaviour of the fresh egg and the older egg can be explained by the dimension of the air bubble within the two eggs. Older eggs have a smaller air bubble because eggs lose air with time.

Finally . . . Some Coffee: But Not to Drink!

Purpose. Familiarity with pH, acid/base indicators, neutralization reaction, hydrolysis and precipitation.

Materials needed. A jar or a bottle, five 100 ml beakers, a small glass, a 2M NaOH solution (be careful using it – can cause severe burns!), a 2M HCl solution (be careful using it – can cause severe burns!), a 0.2% hydroalcoholic phenolphthalein solution, distilled water, $BiCl_3$ 0.1M in 4M HCl solution, 2M KI solution, 3% KIO_3, laboratory gloves and glasses.

Preparation. Fill the jar (or the bottle) with water; add some drops of phenolphthalein solution. The colour of the solution does not change.

Procedure and What's Happening

Beaker n. 1 – red aperitif. Pour 70 ml of water with phenolphthalein (HIn) from the bottle into beaker 1. The colour of the solution turns light red. The reaction is represented by this formula:

$$\underbrace{OH^- + HIn}_{\text{colourless}} \leftrightarrow \underbrace{In^- + H_2O}_{\text{rosé aperitif}}$$

The OH^- displaces the equilibrium to the right: the coloured form becomes prevalent.

Beaker n. 2 – water. Pour the light red solution from beaker n. 1 to beaker n. 2, which contains the HCl solution. The solution become colourless. The reaction is represented by this formula:

$$H^+ + OH^- \leftrightarrow H_2O \text{ (neutralization)}$$

$$\underbrace{H^+ + In^-}_{\text{red}} \leftrightarrow \underbrace{HIn}_{\text{colourless}}$$

The H^+ displaces the equilibrium to the right: the colourless form becomes prevalent.

Beaker n. 3 – red wine. Pour the acid solution from beaker n. 2 to beaker n. 3, which contains NaOH. The colour of the solution turns light red like the colour of wine. The reaction is represented by this formula:

$$OH^- + H^+ \leftrightarrow H_2O \text{ (neutralization)}$$

$$\underbrace{OH^- + HIn}_{\text{colourless}} \leftrightarrow \underbrace{In^- + H_2O}_{\text{rose strawberry}}$$

Beaker n. 4 – almond milk. Pour the solution from beaker n. 3 into beaker n. 4, which contains few drops of $BiCl_3$ acid solution, very slowly. At first the solution is colourless. Add more $BiCl_3$ acid solution and the colour turns white. The reaction is represented by:

$$\text{(hydrolysis) } Bi^{3+} + Cl^- + H_2O \leftrightarrow \underset{\text{(white)}}{BiOCl} + 2\,H^+$$

Beaker n. 5 – orange soda. Pour the "almond milk" into the beaker containing KI. The colour of the solution turns into a colour that resembles orange soda.

Beaker n. 6 – coffee. Pour a little "orange soda" from beaker n. 5 into beaker n. 6, which contains KIO_3 and starch solution. The colour of the solution turns into a dark colour like coffee. The reaction is represented by this formula:

$$6H^+ + IO^-_3 + 5I^- \leftrightarrow 3I_2 + 3H_2O$$

The iodine makes the colour of the starch solution blue. The blue colour combined with the yellow colour makes the colour of the solution dark like coffee.

Characteristics and functions of the anthocyanines. Their name comes from the Greek words νθός (anthos) = flower and κυανός (kyanos) = blue. They are salt glucosides containing three aromatic rings, responsible for the colours red, violet and blue; the purple and mauve colours of many fruits and flowers.

Some of their characteristics include: solubility in water (they colour the water); pH sensitivity (they tend to turn pink/red in an acid environment, green/blue in a basic environment and have no colour in a neutral environment); irreversible oxidation in the air; formation of coloured complexes with metals (iron, copper, aluminium, etc.). The main nutritional characteristics include: anti-oxidative power (larger than vitamins); aperitif and digestive action; they are not absorbed by the intestine mucous membrane. They can be used as pH indicators.

Discovering the Red Onion

Purpose. To show the use of the pigments of a red onion as natural indicators.

Experiment 1. *Materials needed.* A graduated jar (25 ml), a glass jar with a cover, a knife, a red onion, ethylic alcohol, 2M NaOH solution, 2M HCl solution, distilled water, three bars (bar1, bar2 and bar3) with stands, two graduated pipets, a funnel, filter paper, laboratory gloves and glasses.

Preparation and procedure. Cut the red onion into little pieces and put them in the glass jar. Add 25 ml of alcohol and let the onion pieces sit for a few days. Put the filter paper in the funnel and filter the infusion; a light pink filtrate (alcoholic extract of a natural indicator) will be obtained. Mark the jars and pour 2 ml of filtrate (with the pipets) and 3 ml of distilled water into each of them. The colour does not change and the pH is neutral. Pour two drops of 2M HCl solution into the jar1; the colour becomes an intense pink/red. Pour some drops of NaCl 2M into jar2. The colour turns green.

What's happening? The colour variations are related to cyanidine, a pigment found abundantly in red onions (more than anthocyanines). It is a very good natural indicator.

Experiment 2. *Materials needed.* Four pipets, filter paper, NaOH solution 2M, HCl solution 2M, alcoholic extract of a natural indicator (see "Preparation Experiment1"), laboratory gloves and glasses.

Procedure. Using the pipet, put four drops of the alcoholic extract of a natural indicator on the filter paper. Add some water to the first drop of the natural indicator on the filter paper and a drop of HCl 2M to the second one; add a drop NaOH 2M to the third drop on the filter paper; the fourth drop of the natural indicator will be used as a standard of comparison. Observe the variations in colour.

Workshop 2: Food Education

Flora Di Martino

The information and the indexes given here are related to the Italian parameters and nutrition habits and will have to be adapted for Nigeria or other Countries. There are several certified resources on line that could be used according to this purpose, i.e. to understand what kind of food is better for healthy life and how to preserve food in a safe way.

Eat Well and Stay Well

Some Considerations

Using BMI report cards can generate negative reactions in parents; according to a 2010 survey from Trust for America's Health, in U.S. 84 % of parents believe their children are at a healthy weight, even though almost one third of kids are actually overweight or obese.

Purpose. To calculate the Body Mass Index (BMI) and to introduce some basic rules for regular nutrition and a healthy lifestyle.

Materials needed (to be downloaded from web). Panel for the calculation of BMI (Body Mass Index), panel with BMI values, panel of the daily food requirements according to Country, age and lifestyle, a scale, measuring tape, sheets of paper and markers, BMI cards, nutrition facts, ingredients and caloric value for 100 g of each food, "Healthy Weight" Food and Heath cards.

F. Di Martino (✉)
Education and Science Communication projects, Fondazione Idis-Città della Scienza, Naples, Italy
e-mail: dmartino@cittadellascienza.it

A.-M. Bruyas and M. Riccio (eds.), *Science Centres and Science Events*,
DOI 10.1007/978-88-470-2556-1_24, © Springer-Verlag Italia 2013

Table 1 BMI classification

Classification	BMI
Underweight	<18.50
Severe thinness	<16.00
Moderate thinness	16.00–16.99
Mild thinness	17.00–18.49
Normal range	18.50–24.99
Overweight	≥25.00
Pre-obese	25.00–29.99
Obese	≥30.00
Obese class I	30.00–34.99
Obese class II	35.00–39.99
Obese class III	≥40.00

BMI values are age-independent and the same for both sexes. However, BMI may not correspond to the same degree of fatness in different populations due, in part, to different body proportions. The health risks associated with increasing BMI are continuous and the interpretation of BMI grading in relation to risk may differ for different populations. In recent years, there was a growing debate on whether there are possible needs for developing different BMI cut-off points for different ethnic groups due to the increasing evidence that the associations between BMI, percentage of body fat, and body fat distribution differ across populations and therefore, the health risks increase below the cut-off point of 25 kg/m^2 that defines overweight in the current WHO classification.

Procedure

First: weigh and measure the height of one student from the audience. Calculate the Body Mass Index (BMI), which is defined as the weight in kilograms divided by the square of the height in metres (kg/m^2). Discuss the results using the BMI report cards to value if the person is underweight, within a normal range, overweight or pre-obese. Repeat the steps with other students.

Second: ask students their ages, their daily habits (sports, for instance) and other information needed to calculate their daily intake of calories and, thus, their daily food requirements, both variety and required amounts.

Third: build a personal, daily diet. The participants must arrange the list of food groups and relative calories for 100 g of the product ("Healthy Weight" cards).

After the activity is over, the participants can bring their personal, healthy diet home.

Body Mass Index (BMI) is a simple index of weight-for-height that is commonly used to classify underweight, overweight and obesity in adults. It is defined as the weight in kilograms divided by the square of the height in metres (kg/m^2).

For example, an adult who weighs 70 kg and whose height is 1.75 m will have a BMI of 22.9. BMI = 70 (kg)/1.752 (m^2) = 22.9.

Following the table of the International Classification of adult underweight, overweight and obesity according to BMI (Source: Adapted from WHO 1995, WHO 2000 and WHO 2004) (Table 1).

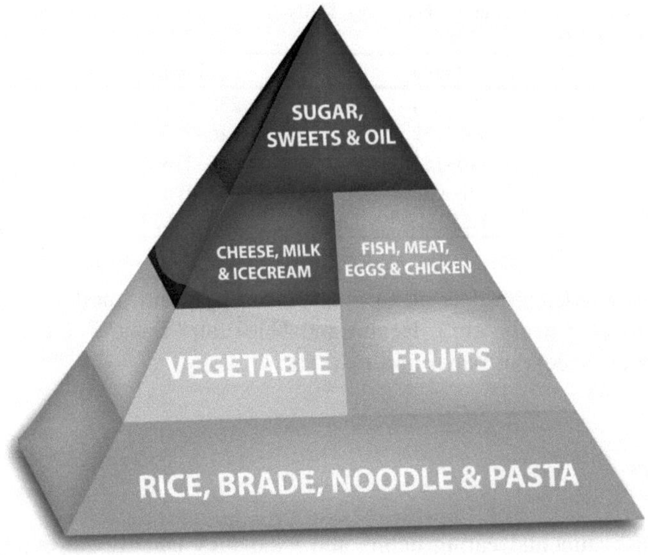

Fig. 1 Food pyramid

Build the Food Pyramid!

Purpose. To introduce the concept of the Food Pyramid. The Food Guide Pyramid is an easy way to create a scheme for a healthy diet. It consists of six food groups and four levels. The general message is that we should eat more servings per day from the lower levels and fewer from the higher ones.

Materials needed. Panel depicting the food pyramid (local foods can be included into the food pyramid, but the basics are recognized at an international level and must be respected); panel of the food groups, Health and Nutrition "Basic Food groups" cards, white index cards size 29.7 × 42, glue, scissors, adhesive tape, makers and pencils.

Procedure. Each person must build his own Food Pyramid. Draw the foods and attach them to the level they are included in beginning with the base of the pyramid and working upwards. Obviously the interaction, the description and the topics of discussion can be adapted according to the public. (I.e. with adults we can talk about the basic rules to eating healthy, calories, food composition, etc.). People can bring their individual food pyramids home (Fig. 1).

Let's Go Grocery Shopping: Foods for a Healthy Breakfast, Lunch, Dinner and Snack

Purpose. To deal what kind of food is necessary for correct nutrition and a healthy lifestyle.

Table 2 Daily food requirements (kcal/day). Daily food requirements (kcal/day) divided in the four principal meals (LARN, 1996)

Meals	3–6 anni	6–11 anni	11–15 anni
Breakfast 15%	188–274	235–341	298–357
Lunch 40%	501–732	626–910	794–952
Snack 10%	125–183	156–228	198–238
Dinner 35%	438–640	548–796	694–833
Total	1,252–1,829	1,565–2,275	1,984–2,380

Materials needed. Models of an empty food pyramid, models of basic foods (i.e. milk, meat, carrots, etc.), Health and Nutrition "Basic Food groups" cards, shopping bags, sheets of paper and markers.

Procedure

First: discuss students' eating habits and what they usually eat every day. Each student receives a shopping bag and goes shopping at the supermarket (boxes of food need to be prepared in advance).

Second: students come back home and organize the food bought in order to make their main meals (breakfast, lunch, dinner and snack).

Third: check if students have built a healthy, daily diet for their three main meals and snack. Discuss about their choices.

After the activity the students can bring the correct list of foods to buy home (Table 2).

How to Preserve Food?

Purpose. To understand the importance of preserving food in order to maintain nutritional properties and avoid threats to our health (i.e. microbial contamination).

Materials needed. Models (or pictures) of basic foods (for example milk, meat, carrot, etc), model (or picture) of a home refrigerator, a larder, an icebox; model of a seasonal nature calendar; Health and Nutrition "Basic Food groups" cards, Health and Nutrition "Food preservation" cards, paper, markers; glue; adhesive tape.

Procedure

First: show the objects representing different foods and different ways to preserve foods (i.e. a home refrigerator, a larder, an icebox) to the students.

Second: ask students to describe how they would arrange the food they have in the different areas of the kitchen. Explain the rules for the correct use, cooking and conservation of food and show them how food can deteriorate. Discuss the three main moments in which the consumer can reduce the risk of microbial contamination: upon purchase, cooking and conservation.

The students can then bring home a handbook on the correct preservation of food.

Notes. If the participants to this lab are families with children, the children can also build a calendar depicting in-season fruits and vegetables.

Extracting DNA

Purpose. To make DNA precipitate and make it visible to our eyes.

Materials needed. Fresh banana piece (about 2 cm cube), mortar and pestle (OR spoon), two beakers (150–250 ml sizes. If no beakers are available, clear plastic drinking cups can be substituted), graduated cylinders (10 and 100 ml), cheesecloth squares, solubilizing (detergent/salt) solution [To prepare 200 ml of this solution (sufficient for ten pieces of banana), dissolve 20 g of non iodized salt into 180 ml of distilled water, then gently mix in 20 ml of liquid detergent (such as a liquid dishwashing detergent), avoiding bubble formation], meat tenderizer solution [To prepare 100 ml of this solution (sufficient for 20 pieces of banana), mix 5 g meat tenderizer into 95 ml distilled water. The liquid will be cloudy], funnel OR rubber band for holding the cheesecloth, cold ethanol (95 %) [6 ml per banana piece used], glass stirring rod, large test tube and test tube rack (or other holder).

Preparation.

- As a first step mash the banana in the solubilizing fluid for a few minutes. Small chunks of banana may remain, since even if many cells fail to be effectively solubilized, the amount of DNA that will be released is vast.
- Filter the cell extract through the cheesecloth slowly. Plan to allow a few minutes for the banana mash to filter. This should produce approximately 10–15 ml of filtrate, which is sufficient for the 6 ml of ethanol in the last step.
- The white masses of DNA will begin to precipitate immediately. If the persons' DNA will be able to spool onto the glass stirring rod, it will do this almost immediately after addition of the ethanol. If the DNA has become too fragmented during the experiment (for example, due to shearing during mixing steps), then you will not be able to wind the DNA. Still, even if the fragments are not long enough to wind onto the rod, they will constitute an impressive amount of soft blobs in the persons' mixtures.

Explain the students that even if the thickness of DNA in imperceptible the length of the DNA molecule and the number of banana cells smashed during the experiment, produce visible threads of DNA.

Key-questions: "What are the parts of DNA?", "Would DNA from a different source look different? Why or why not?", "Why does DNA appear as a viscous material?", "Why would a solution become more viscous after lysis of cells?",

"How long is the DNA in an individual cell and how does the length of DNA compare to the size of a cell?" (see, "DNA" cards).

A Journey Through the Digestive System

Purpose. To show the different steps of the digestive process.

Materials needed. Corn flakes, coffee or tea, a bowl (the mouth), a spoon (the tongue), a strainer, a sieve, a pestle (the teeth), a transparent plastic bag (bag 1 – the stomach), a funnel, a long (1 m $\times$ 5 cm), transparent, flexible, plastic bag (bag 2 – the intestine).

Preparation and procedure. The digestive system is like a perfect machine with communicating tubes and bags. It is specialized in receiving food, in its mechanical crushing and chemical transformation, in the assimilation of the useful substances and in the elimination of non-useful substances.

The journey of food, inside the human body, lasts 24 h. During this journey, food is transformed into simple substances that can be assimilated from the human organism. The journey starts in the mouth where food is mechanically transformed by means of teeth and chemically transformed by means of the tongue and saliva. The result of these transformations is the bolus. It is possible to simulate the transformations that take place in the mouth by putting corn flakes in a bowl. Add some coffee (or tea) to the bowl. Crush the contents of the bowl with the pestle (this simulates the action of the teeth in the mouth) and mix it repeatedly with the spoon (our imaginary tongue). The bolus passes through the oesophagus and arrives in the stomach. Pour the "bolus" into bag 1 (our stomach) and close the bag. Mix the contents of the bag using the hands to simulate the movements of the stomach wall. In the stomach the bolus gets mixed with gastric juices and changes into a more liquid substance called CHYME. It takes about 2 h for time to pass through the intestine. Sieve the contents of bag 1 with the strainer and pour it into bag 2. Squeeze the bag and push the chyme up to the opposite extremity of the bag 2. Discuss the peristaltic movements. Remember that the intestine is 7 m long and is divided into two parts: the small intestine (5 m long) and the large intestine (2 m long), finishing at the anus. The walls of the large intestine are supplied with a lot of little "aspirators" or intestinal villus. The villus permits the useful substances to be absorbed. What is not useful is expelled through the anus in the form of faeces. Pour the chyme into the sieve that separates the non-useful substances (what remains in the sieve) from the useful substances (what passes through the sieve). Give an appropriate form to the materials in the sieve (the faeces!).

Workshop 3: Energy

Alessandra Zanazzi, Anna Porro, and Rossella Parente

Pet Ice Cube

Purpose. To demonstrate how materials have different insulation properties.

Materials needed. Two paper cups of different sizes, ice cubes, insulating materials such as feathers, packing peanuts, styrofoam balls, sand, sawdust, heating pad.

Preparation and procedure. Distribute an ice cube and two cups to four people in the audience and ask them if they know how to keep their ice cube frozen as long as possible. Then ask them to place the ice cube in the smaller cup and then place the smaller cup into the larger cup. Put the different insulating materials between the smaller cup and the larger cup (in one cup put packing peanuts, in the second Styrofoam balls, in the third sand and in the last sawdust). Then ask the participants to describe what is happening in their cups and so which insulating material they think will work best. Place the cups on a heating pad to speed up the melting process.

What's happening? Insulators are materials that slow down energy transfer between the insulated object and its environment. Some of the materials chosen, like foam pellets, will insulate the ice cube better because they do not conduct heat from the heating pad to the ice quickly. Others, like sand, are better conductors of heat, and thus are worse insulators.

Heat Absorption: A Black and White Issue

Purpose. To learn how different colours affect the absorption of heat.

A. Zanazzi • A. Porro (✉) • R. Parente
Education and Science Communication projects, Fondazione Idis-Città della Scienza, Naples, Italy
e-mail: porro@cittadellascienza.it

A.-M. Bruyas and M. Riccio (eds.), *Science Centres and Science Events*,
DOI 10.1007/978-88-470-2556-1_25, © Springer-Verlag Italia 2013

Materials needed. A plastic bottle which has been painted white inside; another bottle which has been painted black inside; two balloons; a heat lamp; a timer.

Preparation and procedure. Place the balloon at the top of each bottle, completely covering its opening. Shine the sun lamp on both bottles and start recording the time. Key-question: "What happens to the balloons? Which one heats up faster?" Record the times for each balloon to stand up straight. *What's Happening*? The lamp heats the air inside both bottles and the air expands. How much and how fast it happens depends from the colour of the bottles. Notice that the balloon on the black bottle inflates faster and more than the balloon on the white bottle. This is because dark colours absorb heat more readily than light colours. Since the black bottle absorbs more heat, the air inside it heats up faster, causing the balloon to inflate more quickly. At a global level, light coloured areas, such as ice and snow, reflect more of the sun's energy. When ice and snow melt, they expose darker coloured soils that absorb heat more readily and cause increased warming of the earth's surface.

The Wool as a Dam

Purpose. To show the behaviour of insulating materials.

Experience 1.

Materials needed. Two Plastic bottles, a wool cloth (a scarf, a sweater, etc.), warm water two thermometers.

Preparation and procedure. Pour the warm water in two bottles. Put the thermometers in each bottle. Wrap the wool clothes around one of the bottles. Read the temperatures on the thermometers and notice that it decreases much more slowly in the bottle wrapped in wool. Discuss the phenomenon and explain why we use wool to protect ourselves from the cold.

Experience 2.

Materials needed. Two glass jars, a wool cloth (a scarf, a sweater, etc.), ice cubes.

Preparation and procedure. Put the ice cubes in both of the jars and wrap the wool cloth around one of them. The ice in the wool melts much more slowly. Discuss the phenomenon and explain why people living in deserts use wool to protect themselves from the heat.

Make a Thermoscope!

Purpose. To learn how to build an instrument to observe increases and decreases of temperature.

Materials needed. A small bottle, plasticine, a transparent straw (or a small tube), food dye (or ink) and warm water.

Preparation and Procedure. Pour out some coloured water in the small bottle and then dip the straw into the water avoiding to touch the bottle bottom. Close the top of the bottle with plasticine and put the jar in warm water (or hold it in one's hands). Observe what it is happening.

What's happening? In the bottle in addition to the water, there is some air. When you heat the bottle, the air expand forcing the water up in the straw. Water is able to expand too, but the expansion of air is much more relevant.

Is It Warm or Cold?

Purpose. To value how the temperature body's receptors respond to the warm and the cold.

Materials needed. Three plastic bottles, warm water, water with ice inside.

Preparation and procedure. Pour the warm water into a bottle. Pour the ice water into a second bottle and the water at room temperature into a third bottle. Have someone grip the bottle with cold water (referred to as "hand 1") in one hand and the bottle with warm water in the other (referred to as "hand 2") for a few minutes. Then put the two bottles down and hold onto the third one (containing the water at room temperature) with both hands. Your hands will perceive two different and opposed thermal sensations – hand 1 feels warm and hand 2 feels cold.

What's happening? The temperature receptors in the body are more sensitive to the last thermal sensation that they felt than to the actual temperature of the object that was touched.

What's Warmer: Wood or Metal?

Purpose. To show that objects at the same temperature can be felt more or less warm according to their conductivity.

Materials needed. A thermometer, a metal object, a wood object, a glass object, a plastic object, warm water, a beaker, a wooden spoon, a metal spoon.

Preparation and procedure. Touch the different objects and decide which one is warmer, colder, etc. Then record the temperature of the objects with a thermometer in order to evaluate the difference between the thermal sensation and the instrumental (objective) valuation of the temperature.

Dip the wooden and metal spoon into warm water for a few minutes. Which spoon gets warmer? Why?

What's happening? Metal has higher thermal conductivity and transfers heat faster than wood.

Ice Cubes Melting on Polystyrene and on Metal

Purpose. To test the thermal conductivity of different materials.

Materials needed. Ice cubes, a polystyrene piece and a metal piece of the same shape and size.

Preparation and procedure. Ask the audience if they know how long it takes an ice cube to melt when it is placed on the two different materials. Put one ice cube on the polystyrene piece and another one on the metal piece. Ask the audience to make hypotheses on what could happen to the ice cubes. Which one melts faster?

What's happening? The ice melts faster on metal than on polystyrene. Metal thermal conductivity is higher than the conductivity of polystyrene. Thus, heat is transferred from metal to ice faster than from polystyrene to ice.

Energy from the Sun: The Solar Oven

Purpose. Through the direct observation of simple phenomena, to build a "solar oven" in order to stimulate children curiosity and invite them to observe, ask questions, play and discover how much energy comes from our Sun.

Materials needed. Thermometer, two cardboard boxes for the inside and outside structure of the solar oven, polystyrene foam supports, aluminium covered cardboard (for the reflector), aluminium foil, aluminium dishes/pans, black colour paint, clear plastic wrap, scissors, a cutter, glue, stapler, black sellotape.

A nice, hot sunny day!

Preparation and Procedure

Phase 1: How the Sun Heats

Discussion with the children/students on the heating properties of the sun.

Begin a discussion about the reasons why the sun is hot. How does the sun radiate heat to the Earth? Children often imagine sun rays as being "physical entities" that transport heat. They do not, in many cases, associate heat to the moon because they say "the moon has no rays. Young children are generally not yet capable of understanding the idea of radiation as energy that moves with no material support.

Another discussion can be begun about the intensity with which the sun transfers heat to different objects (i.e. more heat is felt when one dresses in dark colours rather than light). This can be shown using different coloured metal plates, demonstrating that black absorbs heat better. Metal is a quick heat conductor.

Phase 2: The Solar Oven

This instrument makes it possible to actually use the heat emitted from the sun.

Glue aluminium foil over the insides of the large box, making sure the aluminium stays as smooth as possible and does not crumple. Cover both the inside and

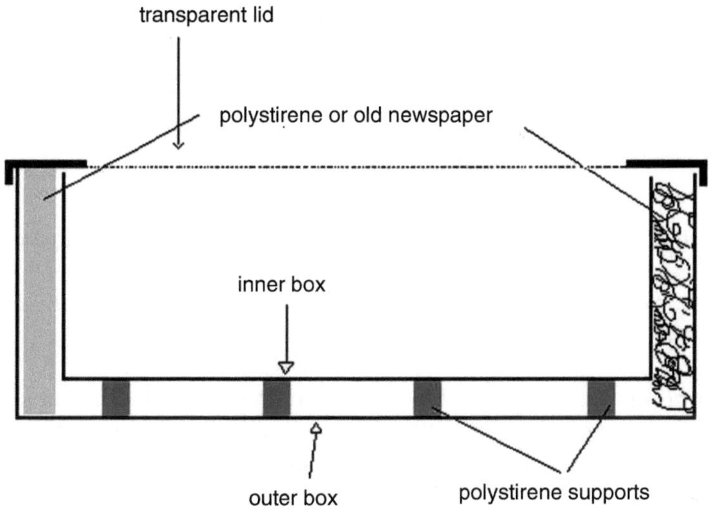

Fig. 1 The solar oven

outside of the smaller box in the same way so that the aluminium reflects the heat and dispersions are limited.

Put the small box inside the large one, filling the space around the two boxes with insulating material (i.e. toilet paper, old newspapers, etc.). Air is also a good insulator and, thus, polystyrene can also be used to support the inner box.

The outer sides of the large box represent the external part of the oven and need to be painted or coloured black in order to absorb as much heat as possible (black works best, as previously discussed). The outer box also has a lid. The external part of the lid has to be coloured black while the internal part has to be covered with aluminium foil. A rectangular hole needs to be made in the lid in order to open a window that will be covered with the plastic wrap and then attached with black sello-tape or something similar. The radiation from the Sun will enter through the window and will be "trapped" inside the oven due to a greenhouse effect. As the lid is held up, more radiation will be concentrated inside the window, as light is reflected from the aluminium foil. The oven and the reflector must also be pointed directly into the sun.

It is also important to use aluminium containers for cooking in the oven (or pots and pans made of good conducting material) that are possibly black (Fig. 1).

Phase 3: Check

We are now ready to cook some food using the oven. We suggest using foods that cook quite quickly and easily, such as apple pieces, an egg, etc. It will take approximately 1 h so it is best to leave the oven in the sun, while you do something else. You will immediately observe that the sun starts evaporating the water inside the food and the vapour condenses in droplets on the surface of the window. You can discuss this with the children.

Experiences on Refraction

Beware: never look at the sun through a magnifying lens! It can damage your retina forever!

Beware: never point laser into someone's eyes! It can cause severe damages to eyes' retina!

Purpose. To show and discuss the phenomena of refraction.

Materials needed. Plexiglass blocks, magnifying lens with different magnifying power, fresnell lens, dropper, little sheets of plexiglass, glass jars of different dimensions, glycerine, beaker, laser, glass slides for microscopy, glass bowl.

Preparation and procedure: *Water as a magnifying lens*. Put some drops of water on a transparent surface and look through the drops. They work as a magnifying lens. The same happens with a glass bowl full of water. In this case, magnified images or upturned images can be observed depending on the distance between the bowl and the eyes.

Missing glass. Put some water in a glass and dip a glass slide into the water. Watch how the slide disappears. A better result can be achieved by melting glycerine in the water.

Diverted light. Point the laser toward a Plexiglass block, you will see the direction of light changing while laser passes from air to plexiglass. *What's happening*? Refraction is the change in direction of a wave due to a change in its speed. This is most commonly seen when a wave passes from one medium to another.

Workshop 4: Environment

Guglielmo Maglio and Riccarda D'Onofrio

Observing Cells

Beware: Never Use Scissors or Scalpel Without Assistance of an Adult!

Purpose. To see and describe the basic structures of cells.

Materials needed. Microscope, onion, tweezers, scalpel, microscopy slides, cover slips, iodine stain, paper and pencils.

Part A. Cell Nucleus and Nucleolus. Onion bulb cells can be used to observe cellular structures called the nucleus and nucleolus.

1. Remove the transparent skin from the inner curve of the onion section. Place this layer onto your microscope slide and add one drop of water.
2. Uncurl or unfold the cell layer. Add one drop of water and one drop of iodine stain to the slide. Add a cover slip.
3. Locate the specimen on scanning power, then change the objective and observe the cells under low power. Notice the brick-like appearance of the cells. The outside layer of each cell is the cell wall. This structure gives plant cells their shape and supports them so the plant can grow taller. Inside the cell wall, but not visible through the microscope, is the cell membrane.
4. Try to locate a small round structure, the nucleus, within each of the cells. The nucleus contains the DNA, which controls cell activity through chemical messages. It is surrounded by the nuclear membrane.

G. Maglio (✉)
Exhibitions Areas and Programs, Science Centre, Fondazione Idis-Città della Scienza, Naples, Italy
e-mail: maglio@cittadellascienza.it

R. D'Onofrio
Education Projects, Fondazione idis-Città della Scienza

A.-M. Bruyas and M. Riccio (eds.), *Science Centres and Science Events*,
DOI 10.1007/978-88-470-2556-1_26, © Springer-Verlag Italia 2013

5. Draw one cell in the space below. Label the cell wall, the cell membrane, the nucleus and the nuclear membrane.
6. Place one nucleus in the centre of your field of view and switch to high power. Inside the nucleus there are two small structures called nucleoli. A nucleolus produces a chemical substance called RNA, which is responsible for carrying the DNA's chemical messages out into the cell.
7. On the diagram that you drew above, sketch and label the nucleoli.
8. Switch to scanning power and remove your slide from the stage. Clean off your slide to use again in Part B.

Part B. Chloroplasts. Another organelle found in the cells of most green plants are the chloroplasts. *Elodea canadensis*, a common water plant, is good for observing these structures. Chloroplasts are responsible for photosynthesis, the process of plants using sunlight to turn carbon dioxide and water into sugar molecules.

1. Take one Elodea leaflet from the growing tip (the end away from the stem). Place it on a slide and add two drops of water. Add a cover slip.
2. Locate your leaflet on scanning power, then switch to low power. Move your specimen so you are looking near the edge of the leaflet. Locate the green, oblong cells. Like all plant cells, they are surrounded by a cell wall with a cell membrane inside.
3. The small green organelles inside each cell are the chloroplasts. Chloroplasts contain the chemical pigment chlorophyll which gives them their green colour. Observe the cells for a minute and you might see the chloroplasts moving around inside the cell.
4. Draw one cell in the space below and label the cell wall, the cell membrane and the chloroplasts.
5. Turn back to scanning power, remove the cover slip and add one drop of iodine to the slide. Replace the cover slip and find the cells again on low, then on high power. Locate the nucleus in one of the cells. Add the nucleus to your sketch and label it.
6. In your Elodea cells, notice that the chloroplasts do not fill up the whole space inside of the cell wall. This is because most of the space is filled with one large vacuole. This vacuole contains water that helps the plant to grow. On your drawing, label the space that is probably the vacuole.
7. Watch this. Switch to scanning power, then have your instructor come over with his magical water dropper and add some salt water to your slide. Watch the cells for minute and see what happens to the material inside the cell wall.
8. Make a sketch in the space below showing what happens to the cells after the salt water is added. You should be able to now see the cell membrane. Label the cell wall, the cell membrane and the chloroplasts on your sketch.
9. Switch to scanning power and remove your slide from the stage. Clean off your slide to use again in Part C.

Part C. Leucoplasts. Potato plants produce and store starch as they grow. The starch is stored as small granules in special organelles called plastids. Plastids can store starch or chemical pigments. Plastids that store starch are called leucoplasts. Since the potato grows underground and is not exposed to sunlight, it will not contain any chloroplasts.

1. Add a drop of water to a slide. Scrape a small piece of potato. Add one drop of iodine to the slide and then add a cover slip.
2. Locate the potato cells using scanning power, then switch to low power. The black areas are starch grains being stored in the leucoplasts. Two or three grains will be surrounded by the cell wall of the potato cell.
3. In the space below, sketch one or two potato cells. Label the cell wall and the starch grains.
4. Switch to scanning power and remove your slide from the stage. Clean off your slide to use again in Part D.

Part D. Animal Cells. Animal cells are the best samples, for observing the cell membrane, since there is no cell wall surrounding the outside of the cell. Animal cells are also good for observing cytoplasm since they do not contain large vacuoles filled with water.

1. Prepare a cheek cell slide as you did in the previous lab. Place a drop of water on a slide. With the flat end of a toothpick, lightly scrape the inside of your cheek to obtain some cells.
2. Stir the end of the toothpick into the water and add one drop of methylene blue. Add a cover slip and locate the cells under scanning power.
3. Centre a group of cheek cells in your field of view, then switch to low power. Find two or three isolated cells and centre them in your view. Switch to high power. You may need to adjust the amount of light by changing the diaphragm setting.
4. You should see the cell membrane surrounding the cell and the nucleus and nuclear membrane. The material between the cell membrane and the nuclear membrane is the cytoplasm.
5. Sketch one cell in the space below. Label the cell membrane, the cytoplasm, the nucleus and the nuclear membrane.
6. Look closely at the cytoplasm. You should notice that the cytoplasm is not smooth, but has a rough appearance. This is because the cytoplasm contains many smaller structures called organelles. These organelles include the endoplasmic reticulum, the ribosomes and the Golgi Apparatus among others.
7. Switch to scanning power and remove your slide from the stage. Clean off your slide and cover slip and place them back in the appropriate containers.

Life in a Water Drop

Purpose. To investigate the living creatures in a drop of pond water under magnification.

Materials needed. Fresh pond water, clear glass jar(s), slides, droppers, hand lenses, microscope, a "Pond Life Identification Kit".

Procedure. Set up the microscopes and invite people to observe microscopic organisms found in pond water using a hand lens and a microscope. Try to identify the organism using a guide book on identification of "Pond Life". Observing these organisms should stimulate discussions about how single and multiple-celled living things might satisfy their needs for food, water, and air. They can do this by comparing the needs of macroscopic organisms to those of microscopic ones.

Baby Bean

Purpose. Using two different preparations it is possible to identify the different parts of a bean and to discover their functions.

Materials needed. About 50 beans, a bowl, water, "the various parts of the bean" cards, transparent box, disks of paper, soil, pot for plants.

Preparation:

1. Put a pair of beans in a transparent box containing a disk of wet paper; repeat for several days, once a day, the same preparation with other beans. Be sure that all the absorbent disks of paper are constantly damp in order to let plants sprout from the beans. After a few days you will see the seeds of the bean in different phases of development.
2. Put the beans in the bowl and cover them with water. Let them soak overnight.

Procedure. Ask the audience if they are able to recognize the different parts of the plants from preparation 1 and then let them discuss their function. Each person, including the explainer, receives a seed treated in preparation 2. The explainer uses his own seed to show the procedure to remove the tegument and to separate the two cotyledons. This must be done by gently placing pressure on the convex side. Then the different parts of the plant can be compared to the images on the cards. Observe the plants in their different phases of development and discuss the function of the different parts of the bean-seeds.

What's happening? A miniature plant can be seen by opening the bean-seed. This is the plant embryo. The functions of each part of the bean-seed are: tegument; the protective covering of the seed; cotyledon, which contains the nourishment (starch) used by the small plant during its growth; epicotile, which produces the leaves; hypocotyls, which will produce the stem; radicle, which will constitute the roots.

Where Are the Roots?

Purpose. To underline the role of roots in a plant's metabolism.

Materials needed. One dish; one musk pad.

Preparation. (1) Let the musk dry for a few days; (2) Put the musk in the dish full of water. After 1 hour you will observe the absence of water in the dish.

Procedure. Key-question: "Where is the water?" Invite a person from the audience to squeeze the musk, revealing how much water squeezes out. *What's Happening*? The musk absorbs the water through the leaves as a sponge. The musk does not have lymphatic vases (the minuscule channels the sap circulates in and where water is contained), but it can absorb a large quantity of water (ten times its weight). The musk can withstand long periods of aridity and helps to keep the soil in the woods damp, favouring the survival of many other plants that suffer from drought.

Coloured Celery

Purpose: To show how water is transported through the stems of the plants and how the changes of turgor pressure influence the stem.

Materials needed: two stems of celery, red food dye, water, two glasses.

Preparation: Make a little cut at the base of the two stems of celery. Put some water in the two glasses; in one of the two add half teaspoon of red dye. Put a stem of celery in the glass with non coloured water and the other in the glass with red coloured water. Let them stand for one night.

Procedure: The stem in the coloured water becomes red (and the leaves assume the same colour). The colour changes as the coloured water climbs up the plant tissue. Key-question: "Are the capillaries thick or thin?" The capillaries inside plants are thin and water climbs up them. Ask some questions about capillary action to begin a discussion with the audience (remember that water can't go higher than 1 m because of the force of gravity). *What's Happening*? Water climbs through the xylematic vases, conductor tubes, which cross the whole length of the stem of the celery. The plants stand erect because their tissues are constituted by water-puffed cells. When a plant doesn't have enough water its cells lose turgor and the stem and leaves bend. The water in the vegetable cells plays a force, called turgor pressure. The capillarity is the result of the combination of cohesion-force between the water molecules and the walls of the tube; it is caused by the difference in attraction of the liquid molecules and the attraction of the liquid molecules to those of the tube. Capillarity is not able to make water rise more than 1 m; two other forces exist to assure a higher ascent of the water in all plants, including the tallest trees – the water's pressure at radical level and leaf transpiration (the loss of water from the leaf surface in the form of steam). Leaf transpiration is particularly important. This is why many water molecules that abandon the leaf are replaced by just as many water molecules recalled from the lower levels because of the cohesion force existing among them.

Friendly Roots

Purpose. To show the importance of the roots for the stability of soil.

Materials needed. A tank, knife, soil with grass, water.

Preparation. Cut the tank lengthwise into two equal parts and fill both of them with loose soil. In one of them sow grass seed and water regularly for 2 weeks until the grass is completely grown. Alternatively, in order to quicken the preparation, the second half of the tank can be immediately filled with a sod (a piece of grassland).

Procedure. Show the two containers to the audience and explain the difference between them. Then invite a person to pour some water in both containers and ask what will happen. In the container with the grass, the water is kept in the soil and only excess water escapes, whereas in the container with loose soil all the water escapes, eroding the soil surface. Key question: "In what way are the observations related to deforestation and floods?" *What's Happening?* This experiment shows how soil erodes. The presence of vegetation prevents the erosion of the surfaces on the mountains slopes, even in case of strong rains. The destruction of forests (deforestation, fires, etc.) can provoke the formation of rivers or mud and water can increase the level of the rivers and streams often provoking real catastrophes.

The Colours of the Leaves

Purpose. To isolate and identify the chlorophyll and carotenoids (coloured pigments of the leaves).

Materials needed. Leaves, small jars; covers for jars, aluminium foil or plastic wrap; rubbing alcohol; paper coffee filters; a shallow pan, hot water, tape, a pen, a plastic knife or spoon, a clock or timer.

Preparation and procedure. Collect large leaves from different trees. Cut the leaves into very small pieces and put them into small jars. Add enough rubbing alcohol into each jar to cover the leaves. Cover the jars with lids, plastic wrap or aluminium foil. Place the jars carefully into a tray containing 3 cm of hot water. Keep the jars in the water for 30 minutes or until the alcohol has become coloured (the darker the better). Twirl the contents of each jar gently about every 5 min. Then cut a long, thin strip of filter paper for each of the jars and label it. Remove jars from water and uncover. Place a strip of filter paper into each jar so that one end is immersed in the alcohol. The alcohol will travel up the paper, carrying the colours with it. After 1 hour or more, the colours will have travelled different distances up the paper as the alcohol evaporates. You will see different shades of green, yellow, orange or red, depending on the type of leaf. Remove the strips of paper, let them dry and then tape them to a piece of plain paper.

What's Happening? The chlorophyll is the most abundant coloured pigment in vegetable cells, responsible for their green coloration. There is a smaller quantity of other pigments in vegetable tissues: the carotenoids (that have colour from the red to the yellow and that are responsible for the colorations of the fruits and the flowers). In this experiment chlorophyll and carotenoids are separate in employing the procedure of the "chromatography on paper". Chromatography means "writing with colours". The different pigments are separated according to their different mixtures. The heaviest mixtures, such as green chlorophyll, are deposited first while lighter mixtures, represented by yellow carotenoids, keep on migrating.

A Flower Blossoms

Purpose. To show the ascent of water by capillarity.

Materials needed. Index cards; scissors; coloured pencils; jar; water.

Preparation. (1) Draw a flower on each index card; (2) Colour and cut out the drawn flowers; (3) Fold the petals of the flowers so that the point of each petal overlaps; (4) Put the "buds" on the water and let them float.

Procedure. Invite the children to build their flowers. Then invite them to put the buds in the water. After a short time the flowers blossom. Ask the children if they know how it happened. Then discuss capillarity by giving some simple examples. *What's happening*? The petals of the flower open due to capillarity. Water slowly seeps through the tiny holes in the fibres of the paper as if they were little tubes. The paper swells and the petals open.

Pollination

Purpose. To discuss and reflect on pollination

Materials needed. Piece of paper

Preparation. Cut paper in small pieces (granules of pollen), pick out a boy (male flower) and a girl (female flower) from the audience. The boy has to place himself in front of the girl at a distance of 2 m. The boy (male flower) will have to swipe at the girl (flower female) with the pollen using only his breath. Afterwards another boy (male flower) and girl (female flower) have to place themselves at the distance of 2 m; in this case, the boy (male flower) gives the small pieces of paper (granules of pollen) to another boy (pollinator bug) who can bring the pollen to the girl (female flower). Which one of the two ways to transport the pollen is more effective?

Water Cycle in a Jar

Purpose. To learn about the water cycle.

Materials needed. Camping stove; two beakers of different sizes; transparent film.

Preparation and Procedure. What is water? Water is a chemical – it is tasteless, colourless and odourless, but is essential to all forms of life on the Earth. Introduce the students to the concept that water can exist in three different states depending on its temperature: Solid – ice; Liquid – liquid water; Gas – steam or clouds;

Water covers more than 70% of the earth's surface. Water is continuously circulating between the Earth's surface and the atmosphere in the water cycle. Set a pot with water on a camping stove and let the water boil. Put some hot water in the smaller beaker and then place it into the larger one. Cover it with transparent film. At the end of the activity you will see that there is also some water in the larger beaker. What happened inside the beaker? Evaporation and condensation can then be discussed with the help of the explainer.

What's happening? This simple experiment will help to understand how water is transferred between the Earth's surface and the atmosphere. Heat from the sun makes water evaporate from the oceans, rivers and lakes forming water vapour. As water vapour rises it cools and starts to condense and becomes water again. This condensation forms clouds. Plants also lose water to the air (called transpiration). As clouds get cooler, water droplets become bigger and bigger. Eventually they fall as rain, snow, sleet, or hail. They can reach the seas and the oceans or the earth's surface directly. In the first case, water can evaporate back into the atmosphere (brief cycle). There are two possibilities in the second case: (1) the water can soak into the soil and some of this underground water gets trapped between rock or clay layers. This is called groundwater; (2) the water flows on the earth's surface. Most of the water flows downhill, eventually flowing back to oceans and lakes (long cycle).

Soil with a Raincoat

Purpose. To evaluate the permeability of different types of soil.

Materials needed: 4 plastic bottles, gauze, adhesive tape, scissors, pitcher, water, gravel, sand, soil, clay.

Preparation. Cut off the bottom part of each of the four bottles and seal the hole on top with gauze. You should now have four funnels. Fill each funnel about 1/3 with the different soils.

Procedure. You can talk about the concept of permeability with the audience and then invite one of the children to pour some water into each funnel. The audience observe and record specific time intervals as the water flows into the soil. At the end of the activity, ask the children to classify each type of soil according to its permeability. Key questions: "Which soil held more water?", "Which soil released more water?", "Where can you see evidence of quick or slow soil absorption in real life? (i.e. sand, a muddy clay-packed road with puddles)". Here there is a scale of permeability: (1) gravel (the most permeable); (2) sand; (3) soil; (4) clay (water doesn't pass). Key questions: "Which soil would be best for a vegetable garden? Why?", "Which soil would be best for a cactus garden? Why?", "Which soil would be best for growing rice? Why?"

What's happening? The larger the spaces between the particles of soil, the more permeable the soil is. Thus gravel, which is made up of little stones, is more permeable because it contains large spaces that water can pass through. Clay is impermeable because it is formed by minute grains.

The majority of terrestrial plants grow better in soil because soil holds the amount of water a plant needs to survive and eliminates excess water that could decay its roots. Terrestrial sand is good for the life of succulent plants (i.e. cacti), whose long, superficial roots are able to absorb water in such terrain by taking in an enormous quantity of rain during the rare moments of precipitation in the areas where these plants are found. Clay, instead, limits air circulation, allowing plants with short, superficial roots to live successfully on this type of soil. Clay is also essential in the formation of natural ponds and for the development of fresh water plants (i.e. water lilies and papyruses).

Greenhouse Effect

Purpose. (1) To show that sunlight (the sun's radiation) can warm the air; (2) To demonstrate that a greenhouse warms up more than the surrounding, "open" atmosphere; (3) To compare a physical greenhouse to the conceptual greenhouse associated with greenhouse gases and climate change.

Materials needed. One half of a clear plastic globe; two small thermometers; two small paper shades used to shield the thermometers from direct sunlight. If working indoors or dealing with cloudy weather, a 100–200 W floodlight in a lamp rated for such wattage. You may need to work near an electrical outlet or use an extension cable. You'll need a solid table or wooden cart to work on. If working on a lightly coloured top, keep dark paper handy to place thermometers and the plastic dome/globe upon. This experiment works best in direct sunlight or indoors under a high wattage light bulb.

Preparation. On a solid table top or on the soil place both thermometers, each under a thermometer shade. Put the globe over one of the thermometers. Let the greenhouse globe and thermometers sit for 10 min or so as you talk to the group about the greenhouse effect. See the notes that follow for topics to engage your audience. Then, take a look at the thermometers. Take note of which one has the highest reading. Ask the group what they think of any observed difference.

What's Happening? The uncovered thermometer, which will display a lower reading, illustrates what life without the greenhouse effect would be like. The plastic globe confines the air warmed by the soil to a limited space, so the air within is warmed and cannot mix with the cooler air surrounding the globe. Although the greenhouse effect is natural (and necessary for life on Earth to exist), it is amplified by human activities like the burning of fossil fuels and deforestation. These add greenhouse gases to the atmosphere, especially carbon dioxide.

Distil Water

Purpose. To learn about the role roots of plants have.

Materials needed. Two glasses, humus, 20 cm of paper towels, water, teaspoon, salt.

Preparation. Put some humus and salt in a glass and mix them together. Pour some water into the glass in order to obtain liquid mud. Twist the paper towel and insert the end into a hole dug in the mud. Dangle the other end inside the second glass and wait for 1 h. Dip your finger into the second glass and taste the water . . . it tastes salty!

What's happening? The paper has absorbed the water contained in the humus. This water climbs up the strip of paper and into the second glass. The water moves through the holes within the paper. The soil can't get through the holes, but the salt absorbed in the water can. This is how roots enable plants to absorb mineral salts from the soil.

Breed the Mould!

Purpose. To determine the conditions in which the mould that produces the antibiotic penicillin grows in. Penicillium is a common green mould that grows on bread and on the surface of fruits, such as oranges and lemons.

Materials needed. Three transparent plastic food bags, two oranges, two lemons, a slice of bread, some absorbent cotton, a dropper, a rubber band, water.

Preparation. (1) Rub the oranges and lemons on the floor and leave them out for one full day. Then place both an orange and a lemon in two plastic bags with a wad of absorbent cotton that has been dampened with water. Seal each bag and place one in a dark place and the other in the refrigerator. Check the bags each day for about 2 weeks. (2) Drop about ten drops of water on the slice of bread and put it in the third bag. Close the bag tightly and keep it in a dark, closed place. Check the bread every day for 5 days. *What's happening*? The orange and lemon kept in the refrigerator seem practically unchanged while the orange and lemon left in a dark place seems to have grown a blue-green beard on its surface, or mould (penicillium). The growth of mould occurs in warm places and is even more encouraged by the presence of dampness. It is much more common for food to become mouldy in the summer months. In the case of the bread, dark spots have appeared on the surface of the slice, producing "mouldy bread". The spores of mould are dispersed in the air and on the surface of bread, once again, encouraged by dampness.

Let's Make a Mould-House!

Purpose. To investigate the good effects of mould. When mould appears on food, most often it means that the food is no longer edible, but some foods (i.e. certain types of cheese) achieve a better taste and enhance their characteristics with the presence of mould!

Materials needed: slices of bread, cheese, fruit, vegetables, an aluminium box, fertile soil from gardening or humus, plastic wrap, a rubber band and a magnifying lens.

Preparation. Fill the aluminium boxes with 2 or 3 cm of soil or humus and place the pieces of food on top. Water the mould-house, cover it with plastic wrap that can be secured with a rubber band. Keep the mould-house in a damp, dark place and check it every day, watering it if necessary. After 3 or 4 days the pieces of food will show some coloured flecks on its surface (blue, grey and white). Remove the plastic wrap. You will smell a terrible odour. Avoid breathing in or sniffing the mould! Using tweezers, remove bits of white mould and place them on a black sheet. Then put some pieces of dark mould on a white sheet. Examine the mould with a magnifying lens or through a microscope.

What's happening? Mould grows well in a warm, damp environment and needs oxygen. Thus, moist food provides an ideal environment for its development. Mould is a type of fungi and, as such, does not get chlorophyll and depends on other organisms to feed. Fungi are very widely spread in the environment in which we live, on our garments and even on our skin.

Oil Suffocates

Purpose. To show the effects of the dispersion of oil on the surface of the water.

Materials needed: a glass, water, drinking straw, oil, mirror.

Preparation. Fill half glass with water. Pour enough oil onto the water to form a thick layer. Dip the drinking straw in the water and set up a mirror so that you can watch yourself. Blow into the straw so that some little bubbles are formed. The little air bubbles rise faster in the water than in the oil. Some little bubbles succeed in overcoming the barrier of oil and to disperse themselves in the atmosphere, while others are trapped in the oil. The oil limits the number of gas exchanges between water and the atmosphere. When some oil is spilled into the ocean, it causes very serious consequences for the equilibrium of the sea ecosystem.

Oil and Birds

Purpose. To observe the effect of oil on the feather of seabirds.

Materials needed: feathers of seabirds, used mineral oil, water.

Preparation. Put the feather in water, dry it and watch it under a microscope. Put the feather in the oil, dry it and put it in the water and watch it under the microscope. Notice the differences.

What's happening. the structure of birds' feathers is unreversibly damaged by oil, which cause the filaments of feathers to stick to each other, affecting the ability to fly.

The Worm Eats My Garbage!

Purpose. Some of the garbage that we produce daily can be recycled to produce compost. You can discuss which are the chemical and biological processes transforming garbage into rich soil.

Materials needed. Microscope, earthworms, centipedes, mould and fermenting food, instructions on how to produce the compost.

Preparation and Procedure. Introduce the topic of garbage. Explain how it is possible to make compost at home. Show the audience the organisms involved in the production of compost. First, show the centipedes. Observe and talk about its behaviour and its mainly anatomical characteristic. Ask someone to touch the animals carefully, so to not to crush them. Afterwards, show the earthworms to the audience. Here you should avoid having the people touch them. Explain that they are very delicate and should not be caressed. The soil where these animals live shows a structure characterized by little spheres that represent the result of the digestion process in animals. Give the audience the information for compost production at home.

Workshop 5: Music and Science

Mario Campanino

Sand in Resonance

Purpose. To demonstrate resonance.

Materials needed. Tambourine, sand (sugar or rice), metal tray, two similar tuning forks.

Preparation and procedure. Put some sand (sugar or rice) on the surface of the tambourine. Strike a metal tray close to the tambourine. The tambourine membrane resonates and the grains of sand jump (leap). Put a tuning fork close to the second one and strike the first one. The second tuning fork starts to vibrate too. Repeat the experience using two equal musical instruments, pluck one of the instrument's strings. You can see the same string of the second instrument start to vibrate.

What's happening? Every elastic body, when you strike it, produces a particular sound. The sound depends on the form, the dimension and the density of the body. If you produce, near an elastic body, a sound similar to the characteristic sound of the body, the waves spread in the air and force the body to vibrate. This phenomenon is called resonance.

Waves Running on the Wire

Purpose. To demonstrate wave propagation on an elastic cord in order to simulate the deformation of the string of a musical instrument.

Materials needed. Three or four elastic cords.

M. Campanino (✉)
Education and Science Communication projects, Fondazione Idis-Città della Scienza, Naples, Italy
e-mail: campanino@cittadellascienza.it

A.-M. Bruyas and M. Riccio (eds.), *Science Centres and Science Events*, 215
DOI 10.1007/978-88-470-2556-1_27, © Springer-Verlag Italia 2013

Preparation and procedure. Two people keep the extremities of the cord. A third person pinches (plucks) the cord close to an extremity, as if it were the string on a guitar. The same form that the cord has at the moment of stimulation (triangle formed between the segment of the cord at rest and the segments in which the cord is divided at the moment of the pinch) spreads along the entire cord, up the extremity and then comes back in a reverse form.

Students can repeat the same experiment in organized groups. Afterwards they can observe the vibration of a string in a string instrument and can try to recognize the same deformation observed on the elastic cord. *What's Happening*? When you pluck the cord, you transmit some energy, which then travels along the cord. When the energy gets to your fingers, it can not go on and goes back along the cord.

Count the Waves

Purpose. To measure the frequency of a wave.

Materials needed. Three or four elastic cords, a chronometer.

Preparation and procedure. Repeat the steps of the "Waves Running on the Wire" activity in order to make participants conscious of wave propagation. Afterwards, repeat the activity with a variant. A new participant must tell the person that has to pinch the cord and stop the activity after 10 s when to start. During this short time, the two people holding the ends of the cord count the number of the impulses that arrive at the end of the cord. Calculate the number of impulses/second; this value is the frequency.

What's Happening? The rate of the energy on the cord is related to the material, the thickness and the tension of the cord. Different speeds produce different vibration frequencies. When you change the tension or the length of the cord, you can observe a different frequency.

The Waves Ring-A-Ring-o'-Roses

Purpose. To simulate the form of a wave.

Materials needed. A ball.

Preparation and procedure. Seven/eight children place themselves in single line. All the children lift up one hand. The first child in line passes a ball (that simulate the spreading stimulus) to the second child in line; the second child passes the ball to the third one and so on. When the ball gets to the last child in line, he must give the ball back to the child in front of him and the stimulus (the ball) runs back along the line.

What's happening? The children simulate the behaviour of waves moving in an elastic body. When the wave meets an obstacle, it goes back.

Big Sound and Little Sound

Materials needed. Musical instruments of different sizes.

Preparation and procedure. Put some musical instruments of different sizes in a row on a table in accordance to size, from the smallest to the biggest. Play the instruments from the biggest one to the smallest and afterwards from the smallest to the biggest one. Key question: "Do sounds go up or do they go down?" Stimulate the growth and reduction of the sounds with body movements (i.e. the explainer can bend down and stand up while he plays the instruments). Invite the children to follow the sounds bending and standing up. Discuss the characteristics of the instruments related to the different sounds.

What's Happening? Musical instruments of different sizes produce sounds of different heights (frequencies). As you have seen in the "COUNT THE WAVES" activity, the smaller the body is, the higher the frequency. With high frequency you have acute sounds, with low frequency, you have low sounds.

Boing

Purpose. To show different kinds of waves (longitudinal, transverse) and their frequency.

Materials needed. Slinky strings.

Preparation and procedure. Hold an end of the slinky allowing it to wave freely (every child can do it). The slinky wave behaves in accordance with the laws of harmonic motion. With the long swaying part of the slinky, the swinging frequency is low; with the short swaying part of the slinky, the swinging frequency is high (students can create different swinging frequencies by evaluating the running time of the wave along the slinky).

Put the slinky on a table and hold it still. Two kinds of waves can be seen: the transverse wave (if you pinch the slinky) and the longitudinal wave (if you compress and then release the slinky). In this last case, attach a piece of adhesive tape to the slinky in order to show that the coils of the slinky do not move. *What's Happening*? When you pluck the slinky in a perpendicular way, you can see the transversal waves (the waves of the musical instrument).When you pluck the slinky in the same direction of its length, you can see the longitudinal waves (like the acoustic waves in the air).

Rule and Laser

Purpose. To show how an elastic body vibrates in order to introduce the concept of wave form.

Materials needed. A metal ruler, a laser pointer, a small mirror, a table, paper board.

Preparation and procedure. Put the ruler in a vertical position, holding one end to the external edge of the table and letting the upper end stand free. Fasten the small mirror to the upper end of the ruler. Turn on the laser and point the ray at the mirror in a way that a bright point is reflected on the paper board. When the ruler oscillates, you can see the bright point moving at the frequency of the ruler.

What's happening? The body's vibration is shown as an oscillating point. If the bright point could leave a trail on paper and the paper could run on the paper board, you could see the form of the wave produced by the oscillating body (see voices and sounds at the computer).

Voices and Sounds at the Computer

Purpose. To show different forms of wave and to introduce the concept of timbre.

Materials needed. A computer with sound editing software, a monitor, a microphone, connection cables.

Preparation and procedure. The software allows for an image to be created in the form of a wave from the different voices and sounds that people make. Discuss the differences between the waves brought about.

What's happening? Every man and every object (musical instruments too) produces a characteristic sound, called "timbre". There are not two people with the same timbre and two violins produced by the same factory never have the same sound! The different forms of waves produced by the computer with different input prove this.

Part VI
Technical Annex 2, Other Examples of Activities and Projects

Science Centre for Peace in the Middle East Project

Anne-Marie Bruyas

During the Ecsite conference in 2002 Città della Scienza, together with the Al Quds University, the Bloomfield Science Museum Jerusalem, Unesco and Ecsite, set up an important project to promote peace and dialogue in the Middle East. They were united by the common goal to foster a peace process that would not rely solely on diplomacy but also on the promotion of good practices designed to achieve a different and brighter future for Palestine. The project consisted of the creation of a Science Centre - an interactive museum of Science and Technology - in the campus of the Palestinian Al-Quds University. It was a fascinating concept that immediately won the support of prestigious scientific institutions such as the Hebrew University of Jerusalem, the University of Naples and the Andrea and Charles Bronfman Philanthropies (ACBP).

This project, whose institutional aim was to promote scientific and technological knowledge in Palestine, set three main goals: (1) to strengthen Palestinian national identity through the creation of a cultural and educational institution that was open to the general public and in which all citizens, in the broadest sense of the word, could identify; (2) to provide support for the educational system and contribute to the development of the region; (3) to promote the dialogue and a greater understanding among the various communities in the Middle East through a collaboration in specific fields such as culture, education, research and technological innovation.

Thanks to this initiative the most prominent institutions involved in the promotion of science within the international community have recognised the important opportunity that this project offers to promote the exchange of information and the achievement of common goals between Israeli and Palestinian culture and their respective communities. In order to implement this project, an international committee was therefore created by the promoting institutions. A work schedule was then drawn up to set up the Science Centre by the following actions:

A.-M. Bruyas (✉)
International Projects and Relations, Fondazione Idis-Città della Scienza, Naples, Italy
e-mail: bruyas@cittadellascienza.it

A.-M. Bruyas and M. Riccio (eds.), *Science Centres and Science Events*,
DOI 10.1007/978-88-470-2556-1_28,

Fig. 1 Palestinian youngsters at the Archimede exhibit (© Al Quds University)

- On the 17th and 18th June 2003, Città della Scienza in Naples hosted its first *international workshop* involving the promoting organisations;
- Between the end of September and the beginning of October 2003, a *feasibility study* was drafted and it was completed in the following months;
- An *exhibition of posters* to promote the project in the international community, financed by Regione Campania and UNESCO, was organised. The exhibition was then presented in 200 cities both in the Euro-Mediterranean region and around the world to coincide with UNESCO'S World Science Day for Peace and Development, held on the 10th November 2003;
- Thanks to financing for the "European Partnership for Peace" programme, a *travelling exhibition* of 500 sq m. on *Mathematics* called "Meet Math" was created by experts from the three communities. Given the general purpose of the exhibition, some ways to encourage inter-cultural dialogue and cooperation were identified in the creation of the exhibition and in the history of Mathematics. The "Meet Math" exhibition was presented firstly at Città della Scienza in Naples, then at the Bloomfield Science Museum in Jerusalem and finally at Al Quds University in East Jerusalem. The exhibition was installed in a large tent, thanks to donations from local institutions in Naples and private donors and it is still on display today (Fig. 1).
- Everything has been accompanied by the training of Palestinian staff with Città della Scienza in Naples and the Bloomfield Science Museum in Jerusalem.

The sharing of the common language of science encouraged and reinforced dialogue and cooperation between the various communities in the area in order to

Fig. 2 Israeli and Palestinian scientists in Meet Math exhibition (© BSMJ)

promote a society based on knowledge, tolerance and mutual respect. Today, thanks to this project, the first science centre in Palestine has now been set up. It is the demonstration that working together can have beneficial effects (Fig. 2).

ACCENT: Action on Climate Change through Engagement, Networks and Tools

Flora Di Martino

ACCENT was a project financed by European Commission under Seventh Research Framework Programme under the program Science in Society and coordinated by Fondazione IDIS-Città della Scienza and involving 14 European and Israeli partners.

Climate change issues are clearly a growing concern for the public today. In recent years, people have received a great deal of information from the media on the causes and consequences of climate change. Phenomena and issues related to the climate change process are complex, and the quantity of information is vast. On the internet, and through other media, citizens and young people have free access to a large variety of in-depth information concerning many aspects of climate change, while they are not often prepared to evaluate such information and to elaborate their own opinions.

Moreover, the involvement of society in climate change issues is fundamental for the development of effective adaptation and mitigation strategies, as well as in the implementation of actions to minimise the impact of climate change. In this realm, there is a need to know which kind of new sustainable lifestyles citizens are willing to adopt in order to reduce the negative impact of climate change.

Fifteen relevant organisations, among science centres, science museums and aquaria, were engaged by the ACCENT project in a coordination action aimed at rationalising their two-way communication practices and at strengthening their collaborations with the scientific community on climate change issues.

ACCENT contributed to a global effort moving the campaign on climate change from the "informative" to the "active" phase through the exchange and dissemination of the best practices, with specific actions that encourage the involvement of citizens in actions and in dialogue (Fig. 1).

F. Di Martino (✉)
Education and Science Communication Projects, Fondazione Idis-Città della Scienza, Naples, Italy
e-mail: dmartino@cittadellascienza.it

A.-M. Bruyas and M. Riccio (eds.), *Science Centres and Science Events*,
DOI 10.1007/978-88-470-2556-1_29,

do you commit to action on climate change?

I do

Fig. 1 Logo of the ACCENT communication campaign

Fig. 2 ACCENT launch event, Experimentarium, Copenhagen

ACCENT took the form of a European participative campaign, where the 15 institutions shared their experience under a common framework, and made interactive and participative communication tools available to a large number of European citizens.

ACCENT assessed the outcomes from the engagement and the participation of the public, and delivered reliable data on the European citizens' opinions on and their perception of climate change issues following the European recommendations.

The project capitalised on its skills in a European sustainable web platform that acted as a "collector", as well as a "disseminator", for any organisations dealing with public engagement in science.

The project was officially launched at the Experimentarium by the Danish Minister of Research, during the COP15 in Copenhagen, in December 2009 (Fig. 2).

The work plan for the achievement of the objectives was based on three main actions:

The first one was the implementation of the I Do campaign at the local level in each partner institution, establishing a calendar of events during the 12 months of the *European Communication Action to the Public* with a list of science demonstrations, theatre performances, participatory games, educational activities, workshops, lectures and debates. Science centres, aquaria and museums are considered as reliable sources for visitors, and they are also recognised as places for debate and dialogue, and they support the development of science rather than just acknowledging it.

With ACCENT, the consortium implemented a communication campaign for the development of a European Action towards society, disseminating the latest results of research related to climate change and engaging people in discussing their exploitation in depth. The activities proposed disseminated some research results and their support for the comprehension of the functions of climate and the earth system, including the ocean and the polar regions, determining the causes of changes in the past and predicting their probable future evolution.

The main audience can be divided into three clusters: schools, families (general public) and interested professionals (scientists, industrials and politicians). The project provided the public with different levels of information, allowing various degrees of depth according to the needs and backgrounds of the public.

The assessment of this process at the European level was done through a specialised web platform addressed to the science centre community of ECSITE. This action also included the dissemination of the project outside the consortium through the community of organisations dealing with public engagement in science and technology.

The second one was the organisation of two Local Citizens Debates (LCD) in each country represented in the project consortium, involving a representative sample of citizens from young people to senior citizens, a well-structured discussion about expectations of climate issues. LCDs are local meetings, involving various actors: scientists from local research institutes, stakeholders, decision makers at the local level, civil society organizations and, of course, citizens. In this way, citizens had the possibility to deal with the local dimension of climate change (at the regional and local levels) in relation to the global dimension.

The goals of the LCDs are the identification of specific needs and the implementation of appropriate processes of agenda-framing, getting people engaged in a real dialogue on the future, where they can express their fears concerning the consequences of climate change, their expectations and their propositions freely, in order to offer useful tools for European decision makers.

Since the ACCENT project aimed to encourage dialogue among citizens, experts and stakeholders on the climate change issues, enhancing the role of science centres in Europe, the best locations for the creation of a faithful and friendly atmosphere were science centres themselves.

The third one was the updating of the web platform www.i-do-climate.eu and the dissemination of the results of all phases of the project. ECSITE was a special "sounding board" for the science centre and museum community. The "web platform for European exchange" disseminated, through the network, the best ACCENT practices in other science centres that were not included in the consortium.

Website url: www.i-do-climate.eu

TIME for Nano, Tools to Increase Mass Engagement for Nanotechnology

Alessandra Drioli

The Time for Nano project finds its context at a time when research on nanotechnologies and nanosciences is widely contributing to the advancement of science. The applications of the results of this research are revolutionising our society in many fields such as medicine, computing, materials science, energy production, and manufacturing, with significant impact. Yet, to the general public, these advances can be invisible or difficult to understand because of the nanoscale and of the complexity of such phenomena.

Studying phenomena and manipulation of matter at the nanoscale and developing nanotechnologies leading to the manufacturing of new products and services is the current approach of European research in this sector. It has a huge potential to improve competitiveness and sustainable development across a wide range of industrial sectors. On the other hand, while understanding and technologies advance, some critical aspects appear in relation to the possible impact of nanotechnologies development on human health and environment. Current knowledge concerning toxicity and eco-toxicity of nanomaterials is still incomplete and many civil society organisations have already pointed out to citizens many possible risks. Moreover, recent public debates on controversial scientific issues – such as genetically modified organisms – have taught us that scientists and society should engage in dialogue sooner rather than later, if we are to build a climate of trust and openness and develop robust policymaking processes.

This was the challenge for nanotechnologies which the project addressed: the fast development of nanotechnologies is raising radically new public policies fostering upstream citizens' participation in the debate about the governance of these emerging issues.

A. Drioli (✉)
Education and Science Communication Projects, Fondazione IDIS-Città della Scienza, Naples, Italy
e-mail: drioli@cittadellascienza.it

A.-M. Bruyas and M. Riccio (eds.), *Science Centres and Science Events*,
DOI 10.1007/978-88-470-2556-1_30, © Springer-Verlag Italia 2013

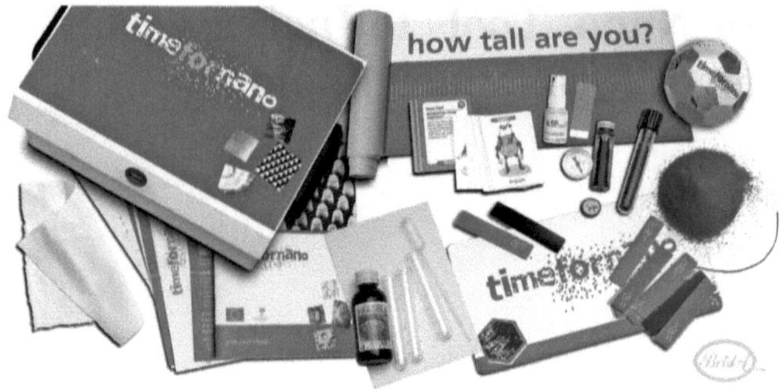

Fig. 1 TIME for Nano educational kit

Time for Nano was designed to ensure that N&N research activities would be made comprehensible to the public and even more should respect fundamental rights and be designed, conducted, implemented, disseminated and used in the interest of the well-being of individuals and society.

Time for Nano developed and spread creative new ways of engaging young Europeans on the topic of nanotechnology. The field of nanoscale research, engineering and technology has always posed significant challenges to science communicators, and this project demonstrated a number of successful methods of engaging the general public, and particularly young people, on the associated benefits and risks.

The project and its tools had a significantly wide impact: The **Nanokit**, a box full of specific activities to be carried out using real nanomaterials (Fig. 1); and the annual **online video contest**, a YouTube-based European competition inspiring young people to create their own films to explore themes of nanotechnology. The **Nanokit** was developed, produced and distributed during the first phase of the project, and the second phase saw its continued use. The kit continued to be used in the science centres or in classrooms (Fig. 2).

In this period, 20,000 people attended **Nanoday** events in the science centres that used the nanokit as well as to encourage participation to the edition of the **online video contest,** which received more than 200 entries. **Nanoday** events, which took place across Europe, engaged young people face-to-face with science communicators and researchers (Fig. 3).

The organisation of 74 **multipliers' training sessions** was crucial in developing this community, resulting in 953 multipliers fully trained. This community grew significantly in the second half of the project, through partners' national and regional relationships with science centres, science festivals, and schools; and particularly through Ecsite and its large community of science centres and science museums all over Europe. The web platform, promoting the project, hosting resources and acting as the hub of the community of users of the project, received 20,000 unique hits, sustaining the community online. Links with other European projects were also crucial in this community development and the online.

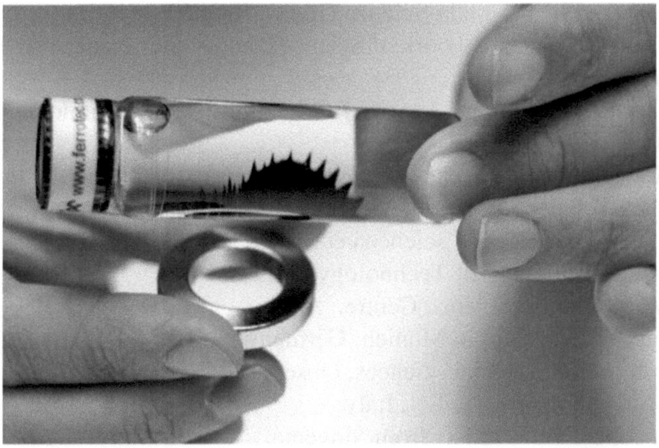

Fig. 2 TIME for Nano educational kit

Fig. 3 Workshop at Heureka, Finland (© Heureka)

From a pedagogical perspective, the two central methodologies of the project were **enquiry-based learning** and **debate**. Since the project made use of real materials related to nanotechnology, as part of the Nanokit, enquiry-based learning allowed the young people to engage with the objects in an ideal way. As for debate, the project tackled five central controversial topics: Health, Privacy, Environment, Socioeconomic divide and Improvement. Each controversial topics has been associated with a video made by BridA, a collective of artists, so as to encourage creativity.

The project was funded by the European Commission under the Seven Framework program. Thirteen partners from ten different countries, a well-balanced representation of countries from all over Europe, worked on the project:

- Fondazione IDIS-Città della Scienza (IDIS), Italy.
- ECSITE, the European network of Science centres and science museum, Ecsite, Belgium.
- Ciência Viva, the National Agency for Scientific and Technological Culture in Portugal.
- CCSTI La Casemate, a French science centre based in Grenoble, France.
- The Turkey Science Centres Foundation.
- Technopolis®, the Flemish science centre, Belgium.
- The Warsaw University of Technology, Poland.
- Heureka, the Finnish Science Centre.
- The Deutsches Museum in Munich, Germany.
- A research group in Social sciences, Observa, Italy.
- A publishing company, CUEN, Italy.
- BridA, a collective of artists from Slovenia.
- The Association for Science and Discovery Centres, United Kingdom.

Project website: www.timefornano.eu

References

Amodio L., Buffardi A., Savonardo L.: La cultura interattiva, comunicazione scientifica, musei, science centre. Pomigliano d'Arco: Oxiana (2005).

Amodio L.: Interagire, parlare, agire. Una nuova "interattività nei musei scientifici. In: N. Pitrelli N., Sturloni G. (eds.): Governare la scienza della società del rischio. Atti del IV Convegno Nazionale sulla Comunicazione della Scienza. Milan: Polimetrica International Scientific Publisher (2006).

Antinucci F.: Comunicare nel museo. Rome-Bari: Laterza (2004).

Bandelli A., Konijn E., Willems W.: The need for public participation in the governance of science centers DOI:dx.doi.org. Museum Management and Curatorship **24**(2), 89–104. Amsterdam: VU University (2009).

Bandelli A., Konijn E.: An experimental approach to strengthen the role of science centers in the governance of science. In: Marstine J.C. (ed.): The Routledge Companion to Museum Ethics, 164–173. London: Routledge (2011).

Beretta M.: Storia materiale della scienza. Milan: Mondadori (2002).

Bertuglia C.S., Infusino S., Stanghellini A.: Il museo educativo. Milan: Franco Angeli (2004).

Bruman R. et al.: Exploratorium Cookbook I . San Francisco: Exploratorium (1976–1991).

Bucchi M.: Scegliere il mondo che vogliamo. Bologna: Il Mulino (2006).

Castelfranchi Y., Pitrelli N.: Come si comunica la scienza?. Rome-Bari: Laterza (2007).

Cerroni A.: Scienza e società della conoscenza. Torino: UTET (2007).

Durant J.: Museums and the Public Understanding of Science. London: NMSI Trading Ltd Science Museum (1992).

Greco P.: La Città della Scienza. Storia di un sogno a Bagnoli. Turin: Bollati Boringhieri (2006).

Greco P., Silvestrini V.: The never-ending resource. Naples: CUEN (2010). La risorsa infinita. Per una società democratica della conoscenza. Rome: Editori Riuniti university press (2009).

Hipschman R. et al.: Exploratorium Cookbook II. San Francisco: Exploratorium (1980–1983).

Hipschman R. et al.: Exploratorium Cookbook III. San Francisco: Exploratorium (1987–1993).

Koster E., Falk J.: Maximizing the external value of museums. Curator **50**(2), 191–196 (2007).

McLean K.: Planning for people in museum exhibitions. Washington D.C.: Association of Science-Technology Centres (1993).

Deridder J., Nisin R., Sabri M.L.: Handbook: organise a project for public understanding of science. Institut de Recherche pour le Développement (IRD) in partnership with the association Planète Sciences: http://www.latitudesciences.ird.fr (2007).

Migueres M.E., Sabri M.L.: Handbook for Science Educators. Institut de Recherche pour le Développement (IRD) in partnership with the association Planète Sciences: http://www.latitudesciences.ird.fr (2007).

Migueres M.E., Nisin R., Sabri M.L.: Handbook: create and run a science centre. Institut de recherche pour le développement (IRD) in partnership with the association Planète Science: http://www.latitudesciences.ird.fr (2009).

A.-M. Bruyas and M. Riccio (eds.), *Science Centres and Science Events*,
DOI 10.1007/978-88-470-2556-1, © Springer-Verlag Italia 2013

Nowotny H., Scott P., Gibbons M.: Re-thinking science: knowledge and the public in an age of uncertainty. Cambridge: Polity Press (2001).

Pascolini A.: Efficacia delle mostre per la promozione della cultura scientifica e condizioni per il loro successo. In: Pitrelli N., Sturloni G. (eds.): La stella nova. Atti del III Convegno Nazionale sulla Comunicazione della Scienza. Milan: Polimetrica International Scientific Publisher (2005).

Rodari P.: Il visitatore al potere. Il dibattito contemporaneo sul ruolo dei musei della scienza. In: Pitrelli N., Sturloni G. (eds.): La stella nova. Atti del III Convegno Nazionale sulla Comunicazione della Scienza. Milan: Polimetrica International Scientific Publisher (2005).

Rodari P., Merzagora M.: La scienza in mostra. Milan: Bruno Mondatori (2007).

Stocklmayer S. M., Gore M.M., Bryant C. (eds.): Science communication in theory and practice. Boston: Kluwer Academic Publishers (2001).

Wagensberg J.: The Total Museum. A tool for social change. Provocative Paper, 4th Science Centre World Congress, Rio de Janeiro 10–14 April 2005 (2005).

Ziman J.: Real science. Cambridge University Press (2002).

AA.VV.: The manual of museum exhibition. Oxford: Altamira Press (2002).

MIX
Papier aus verantwortungsvollen Quellen
Paper from responsible sources
FSC® C105338

If you have any concerns about our products,
you can contact us on
ProductSafety@springernature.com

In case Publisher is established outside the EU,
the EU authorized representative is:
Springer Nature Customer Service Center GmbH
Europaplatz 3, 69115 Heidelberg, Germany

Printed by Libri Plureos GmbH
in Hamburg, Germany